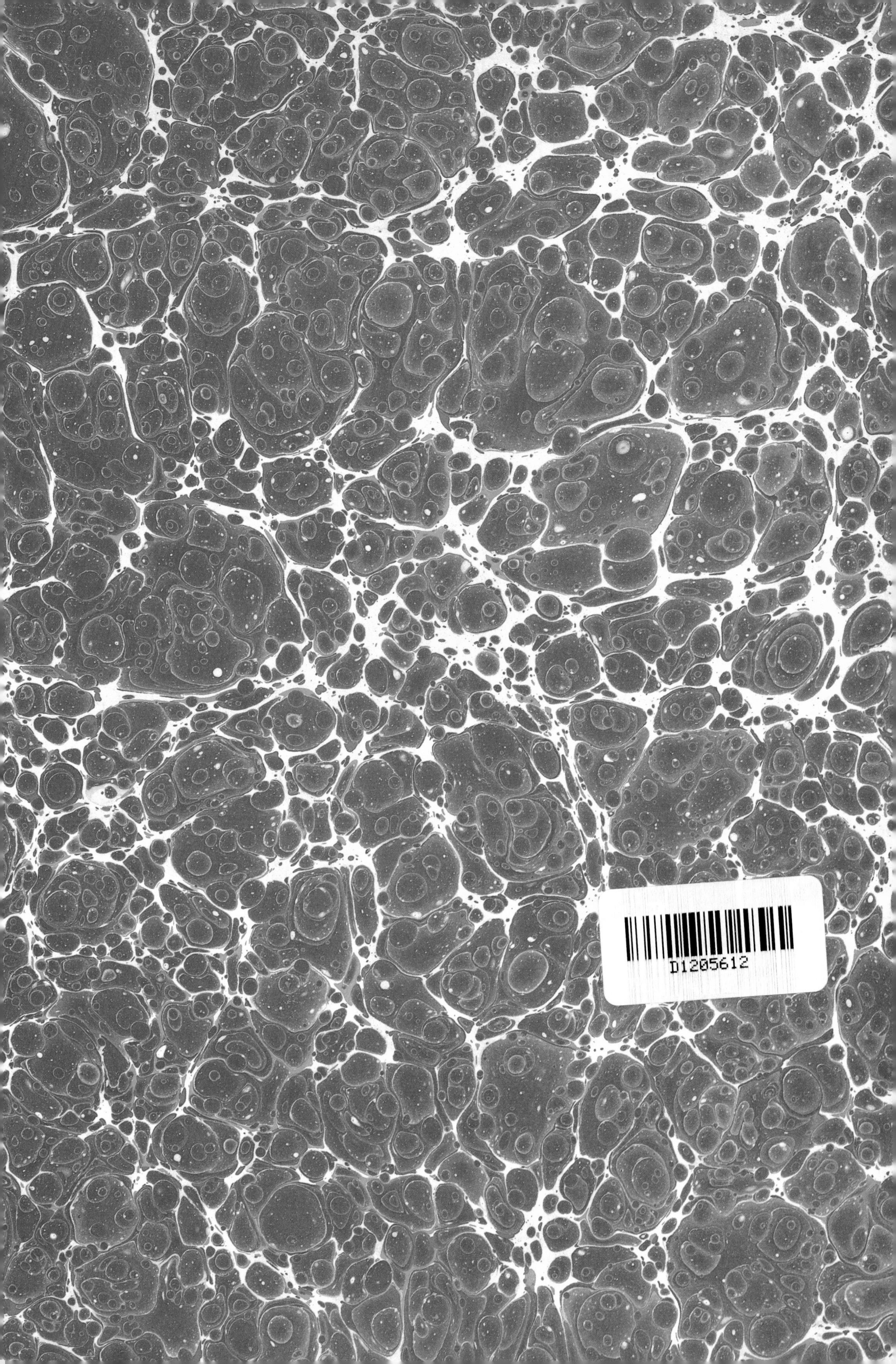
D1205612

PRINCE of DOLLMAKERS

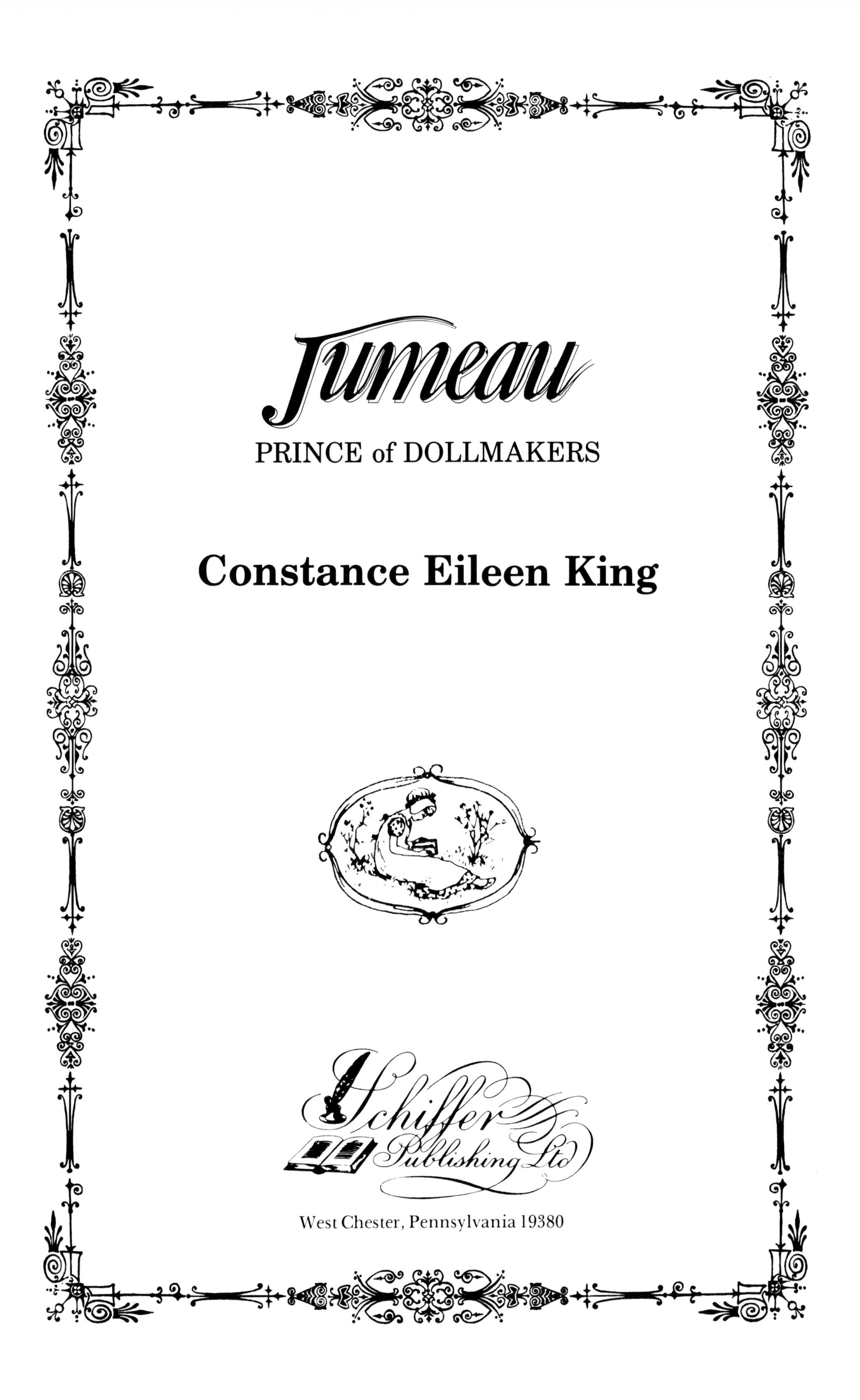

Jumeau

PRINCE of DOLLMAKERS

Constance Eileen King

Schiffer Publishing Ltd

West Chester, Pennsylvania 19380

Library of Congress Catalog Number: 87-62444

Photographs by Acanthus.
Printed in the United States of America.
ISBN: 0-88740-115-5
Published by Schiffer Publishing Ltd.
1469 Morstein Road, West Chester, Pennsylvania 19380

This book may be purchased from the publisher.
Please include $2.00 postage.
Try your bookstore first.

Acknowledgements

This book is intended to present the work and history of the Jumeau factory with a completeness impossible for an individual collector or museum. Though one person or institution might well own as many examples, space, lighting and labelling of each piece would make the task extremely difficult. These examples were chosen not simply from among the most expensive dolls, but also from the cheaper, open-mouth versions which the majority of collectors purchase first. On the back of the jacket for instance, is a tiny stiff-legged example marked "J" and surely among the cheapest dolls produced by the firm. At the other extreme are fine portrait dolls and automata. Some were photographed as detailed portraits so that every aspect of the faces can be studied. This feature should be especially helpful to the many people who now make reproductions and find small photographs very difficult to use as guides for painting.

I would especially like to thank the collectors and dealers who allowed me to examine and photograph their dolls; Abla Odell, who loves variety but insists on complete originality, Rachel Karslake, who thrives on the restoration of damaged specimens and Dorothy Brooke, who loves her Jumeaux because of their great beauty. At the home of Emma Berry, I saw how a group of exquisite examples can transform a collector's lifestyle. Another enthusiast, Polly Edge, was busy setting up a new museum, using her Jumeau, with its large trunk of clothes, as a central feature. My thanks also to the group of enthusiastic ladies who run a doll shop in Hungerford but seem to sell more to each other than to their customers: Dawn Herrington, Pat Birkett and Brenda Foster. David Barrington, a dealer who has handled some of the finest Jumeaux, and Lilian Middleton, another dealer and collector whose skill as a couturier-dressmaker has inspired her to re- create Jumeaux costumes. My thanks also to a group of collectors from the Essex doll club, Lilian Wood, Denise St Clair and Terice Tipper, whose dolls exhibited their wide spectrum of interest, from costume to social history. Betty Woods kindly added her Jumeau to an interesting group and Bunny Campione of Sotheby's, London, endured photography during a busy viewing. Also to Pamela Brown Sherer of Sotheby's, New York who supplied two photographs of fine automata. Anna Marrett of Phillips, London, also supplied black and white photographs, as did Olivia Bristol of Christie's, South Kensington. Without the co-operation of all these people it would have been necessary to use museum or picture library examples, always a little unsatisfactory, as we all like to see dolls that are attainable.

All the border patterns and edgings used in this book are taken from Jumeau's design for the 1886 lottery.

Preface

The dolls made by the Jumeau factory in France have attracted many generations of collectors. The quality of the bisque heads, the cleverly contrived, couturier-designed costumes and the skill of the body construction all combined to ensure that they stayed in the forefront of popularity. When first produced, Jumeaux were comparatively expensive, a position which they have held to the present.

Any object that is almost immediately linked to expense seems to arouse an adverse reaction. Perhaps these are bank account dolls, appealing mainly to investors. We turn away, but then look again into the doll's face and become devotees.

When photography for this book began, I was concerned that some of the 1890's Jumeaux faces would appear too similar, but as work progressed I became more aware of the individuality of each, dependent on its painting or costuming. One had the most exquisitely painted lips, another spectacularly large eyes . . . Each seemed to come alive on examination and stand quite apart, even from others made from the same mould. It became easy to see why collectors are so dazzled by Jumeau that they concentrate almost exclusively on the firm's work.

The study of this company, through exhibition reports, patent specifications and contemporary accounts, made me appreciate more completely the dolls in my own collection. The early portrait type with slightly uneven eyes seemed to increase in appeal with knowledge of the problems associated with this section of the manufacturing process, while a Parisienne also gained attractiveness because of her early date. Through hours spent in photography others absorbed my interest because of their association with proud owners. One thought of her dolls as real works of art, another used them as a guide for making reproductions, while others watched over them like anxious mothers, arranging their ringlets and straightening their skirts. The modern dollmakers were most inspired by the painting of the Jumeaux heads, using the dolls as a model upon which to base their own work.

There was, however one feature all had in common, a real love of the individual dolls that was quite different in character to that seen, for instance, among collectors of silver or porcelain. The dolls, because of their clothes, their moving eyes, the occasional speaking mechanism and their good articulation, have a sometimes almost disturbing affinity with living things, which encourages a commitment from the owner or viewer far deeper than that engendered by a fine paperweight or statuette.

These Pampered Children

Foreign visitors to France in the second half of the nineteenth century were invariably surprised at the number of children always visible, especially in Paris, where there seemed to be more than in any city in Europe. French infants were not kept out of adults' sight in the nursery or in the care of nannies, as in England, and consequently these well dressed little creatures were to be seen regularly in many of the most fashionable places. The native love of strolling in the parks or the boulevards, itself encouraged a display of the person, and even in weather that in any other country would have been considered too cold for socialising out of doors, the French would continue to parade, often accompanied by young babies, so flimsily dressed that German women in particular feared for their safety. In order to amuse the children on these regular public excursions there were gaily painted swings and roundabouts situated in many of the most fashionable walking areas, and these were full, all day long, with the beautifully dressed, laughing children that were an essential feature of Paris. Repeatedly travellers commented that French children seemed to be the most loved and happy infants in the whole world.

It was for these adored children that manufacturers such as Jumeau created the most extravagant dolls with costumes and accessories that mirrored the elegance of the adults who, by their perfection, made the streets of Paris a continual, lavish spectacle which the poorer people seemed to enjoy rather than envy. Often in the busy parks, small goat-carriages were to be seen with their juvenile occupants taking the air with sophisticated assurance and displaying their beautiful dolls, whose costumes competed with those carried by other girls. Curiously, these middle and upper class infants, so pampered and feted, were also subjected to moments of harsh reality, and it was not unusual for well dressed boys and girls to be seen walking around the railings of the Paris morgue (where the suicides and drowned bodies were laid out naked for identification) in the care of their mother, or even a nurse, who was using this grim place for an assignation with a young subaltern. The bourgeois child's life was therefore simultaneously one of great luxury and absolute reality.

Even in their homes, the young were not completely separated from the truth of contemporary life, as many families lived in the prestige apartments of a building, while in the cellars and garrets existed people from the very poorest sections of French society, whose lives were a constant battle for the scraps that meant survival for another day. The Parisian bourgeois girl in the middle years of the century was therefore adored and pampered, and yet also made aware of how unpleasant life could be, should she wander in any way from the security of her immediate family. Though she was loved and protected with enthusiasm, this was a care that also bound her to parents, cousins and grandparents, from whose aspirations and social environment she could never become free, as, even after an early, arranged marriage, it would be the demands of the complete family that would be pre-eminent. While English and American children might have envied the clothes and the possessions of the *jeune fille,* they would rarely have wished to change places with individuals who were virtual prisoners of convention and family.

It was only in the early years of the twentieth century that French girls began to enjoy the freedom taken for granted by the young of other nations: the freedom to meet children of other families without formality, to dress as they wished or to visit art galleries, shops and restaurants in the company of people of their own age. In the nineteenth century, the girl from infancy was expected to behave exactly as her parents commanded, and even six or seven year olds would have considered it quite improper to talk with a child to whom they had not been formally introduced. This necessity for an adult seal of approval on any childish relationship was in fact extremely practical, as it protected the family wealth that might have been put at risk in later years because of an undesirable acquaintance in childhood with a boy from a less perfectly matched background.

The extravagant French doll, criticised so sanctimoniously by foreign visitors familiar with the plainer wax or wooden toys, is seen, in the context of the girl's life style, as the logical outcome of the need to provide her with some outlet for imagination and energy which, in a less oppressive environment, would have been served by time spent with brothers and sisters and in much freer association with people of her own age. It was for heavily protected girls such as these that the Parisian dollmakers used all their skill to create figures that engaged the attention and made the long hours spent in the drawing room with mama as sole companion more enjoyable. The mid-nineteenth century dolls were costumed in imitation of the most elegantly dressed adults, so that, while playing, the little girl could imagine herself grown up. Though these fine toys were essentially French in character, a great many were exported, and if the young of America or Bavaria found them quite unlike the women they saw around them, this, in a way, added to their allure as products of that fashionable, fast city, Paris.

The firm of Jumeau was typical of hundreds of family businesses working from small, often single room factories, in the city and its suburbs. Several families or parterships frequently occupied the same mean building, so that the narrow streets were scenes of frantic activity, all related to the production of the charming frivolous items, mainly of a feminine character, which the French were so skilled in creating. Foreign women were in awe of the way in which even the poorest little dressmaker, occupying a corner of a box-like room in one of these buildings, could create a ravishingly expensive effect from a few cheap scraps of material, and it was upon such natural native skill that the success of the doll industry depended. As the basic dolls, made of papier mâché, were, in the 1840's, of an almost universal quality and appearance, the French relied for sales upon excellence in needlework and design. As it was impossible for German or English makers to compete with Parisian skill in creating an enchanting costume, the dolls achieved world fame as luxury items, completely dependent upon decoration rather than basic manufacture, a facet of French industry that was much in evidence in other fields, such as the making of furniture and clocks.

Contemporary accounts suggest that there was comparatively little dollmaking in France in the first half of the nineteenth century. Those produced were of moulded papier mâché and so similar to those of Sonneberg origin that they are impossible to differentiate unless the original costumes are still worn. The French doll sellers also relied heavily on the use of German-made heads which were mounted on shaped leather bodies, often made in France,which further complicates attribition, as no marked dolls have been discovered. Identical dolls are therefore sometimes found in museums and collections attributed, at the whim of the cataloguer, to a French or German source. The only certain fact is that, at the period of manufacture, those dressed in France were considered the finest in Europe. As glazed porcelain slowly replaced papier mâché in Germany as a cheap and effective medium for shoulder heads, these products were also purchased by the Parisian assemblers. It is again impossible to be categorical about the dependence on German-made heads in the 1840's and 50's, and some of the pink tinted heads whose modelling seems so different to the usual German type are often thought to be of French origin. As there is little documentation and none are adequately marked for positive attribution, no firm conclusions can be drawn, except that the number made must have been low.

Some light is thrown on this problem by Natalis Rondot, a member of the French Imperial Jury at the Great Exhibition of 1851. Rondot, an economist, whose information seems reliable, was also at the Great Exhibition as a delegate from the Lyons Chamber of Commerce and, writing in 1855 stated that "A few dolls' heads of porcelain are made in Paris, in the white they cost 9fr. the dozen and one has to pay 12 fr. for painting them". Possibly of even greater interest is his comment that the busts from Germany were quite well made with "The back part of the head cut away, the gap filled with cork to fix the hair on", establishing a date much earlier than is generally accepted for china dolls with cork pates. In general, Rondot's account provides only extra proof of the almost complete dependence of the French dollmakers in the 1850's on German suppliers of heads of both papier mâché and the less popular china.

The importance of the Jumeau factory's contribution to the development of the doll industry in France was that it was not prepared to rely completely on foreign imports, but worked consistently towards a doll that was made in its entirety to its own

BÉBÉ-JUMEAU

DIPLOME D'HONNEUR

RÉCOMPENSE UNIQUE DÉCERNÉE JUSQU'A CE JOUR DANS TOUT LE JOUET FRANÇAIS

LE BÉBÉ-JUMEAU

Jolie tête en biscuit, cheveux longs et soyeux, yeux d'émail. corps incassable se prêtant à tous les mouvements, prenant toutes les positions, chapeau à plumes à la dernière mode, robe de laine ou de soie inventée par le premier couturier parisien, souliers de chevreau doré, bas roses ou bleus à jour, gants de Suède, collier au cou, pendants aux oreilles, toujours souriant, adorablement mignon, tel est le BÉBÉ-JUMEAU.

Rien n'a été négligé pour faire de ce bébé, le plus beau, le plus complet, le plus amusant des jouets. C'est le cadeau par excellence pour nos fillettes, petites et grandes, avec lui on ne s'ennuie jamais ; il s'habille, se déshabille, lève les bras, s'agenouille ou s'assoit, il est d'une obéissance à servir de modèle à ses petites mamans,

Il ne se casse pas le BÉBÉ-JUMEAU : il est éternel ; quand il a usé sa belle robe brochée on lui en refait une autre et en voilà pour longtemps.

Il se tient droit sur ses jambes, solide et tendant les bras vers qui l'appelle ; il ne lui manque que la parole, et encore ! — Si l'on veut on lui fait dire : Papa ! Maman ! à haute et intelligible voix. Afin qu'on soit fixé sur son identité et qu'on n'achète pas de mauvaises imitations, le BÉBÉ-JUMEAU à un passeport en regle ; il est paraphé, marqué ; il a *son nom* sur le *corps* et quand il est habillé, sur le *bras gauche*. Il est quelqu'un — et la loi veut qu'on le respecte.

Le BÉBÉ-JUMEAU a des amis dans tous les mondes et sait se plier aux exigences de la société, car il est bien élevé le BÉBE-JUMEAU.

Il voyage : Londres, Vienne, New-York, Milan Madrid, toutes les capitales du monde civilisé ont reçu sa visite, et dans sa mère-patrie il a su s'attirer la sympathie générale. Paris l'idolâtre ; Lille, Roubaix, Nantes, Nancy, Strasbourg, Metz et Mulhouse l'ont adopté et lui font tous les ans, à cette époque-ci, la plus merveilleuse réception du monde !

Il faut que toutes nos petites Françaises aient un BÉBÉ-JUMEAU pour étrennes, cette année. Le BÉBÉ-JUMEAU est la vraie poupée française, la petite Parisienne, gracieuse et jolie au possible, qui apporte la joie et la satisfaction des plus beaux rêves de Noël à tous les enfants de France

Heureux l'enfant qui peut le posséder.

N° 24686

Ce numéro donne droit au tirage d'un superbe BÉBÉ habillé, qui sera tiré le 20 décembre 1886.

Le numéro gagnant sera publié dans le PETIT JOURNAL.

Left: The 1886 Jumeau Lottery card. The winning number was published in Le Petit Journal on the 20th December 1886. (A translation is provided in the text).

Right: Only in a contemporary photograph such as this do we see the Bébé Jumeau in its pristine silk costume complete with the much discussed arm band. This photograph was printed on one side of the 1886 Jumeau lottery card. Above the dolls is a framed photograph of the large Jumeau factory.
Courtesy Phillips, London.

specifications. Even when the unglazed bisque-headed dolls, that were to replace the porcelains, were made in some number in Paris in the last quarter of the nineteenth century, other producers did not manufacture their own, but sent their designs to specialist factories, where the heads, made to their requirements, were fired and usually decorated. The difference of quality between these heads that were marked with the various firms' names or initials was therefore dependent on exact co-operation between the porcelain factory and the dollmaker. The house of Jumeau was not prepared to rely on other factories for the success of its product, and by amalgamating all the necessary processes of manufacture, it at once became a leader in this section of French industry, inspiring the praise of manufacturers such as Péan Frère.

The firm's founder, Pierre François, began to manufacture dolls in the 1840's, his first listing in the Paris Almanac of Commerce relating to the year 1842, when he was described as being in partnership with Belton as a maker of leather dolls and dolls that were costumed. The information given by the firm to the Jury of the Vienna Exhibition in 1873 states, however, that dollmaking began in 1843 and the Jumeau "Notice" of 1885 also gives the later date, suggesting that the information given to public bodies by Emile Jumeau was not always correct or, more interestingly, perhaps indicating the time when his father gained real control of the small factory from his partner. Belton has remained a shadowy figure and is only known because of his short association with

Jumeau during this early period and he separated completely from Pierre François sometime in 1846, the partnership having been awarded an Honourable Mention at the 1844 Paris Exposition of Industry.

The early years

A great deal of the early history of the Jumeau factory has to be traced through its participation in the exhibitions that had been the centre of French commercial life since the late eighteenth century, when the Marquis d'Avèze originated a series of fêtes at St. Cloud, at which manufacturers were invited to display their goods, a series that culminated in the 1797 show, held in the Champ de Mars. The second Exposition took place under the Consulate in 1801 and was followed by an intermittent series, one of the largest being that of 1849 that opened, like several others, in an incomplete state. The importance both of exhibiting at such events and obtaining awards is usually much exaggerated by writers on dolls, though they were of use in showing the public the latest advances in design. Obviously the greatest importance of these events was in the mechanical and agricultural sections and the inventions of Jacquard, Aubert and Vaucanson were all brought first to public attention at such exhibitions.

The 1844 Exposition, the last at which Louis Philippe presided, was the most splendid arrangement of French industrial products ever assembled and it was obvious that, though behind Germany and Britain in the first half of the century, France was making better use of all the advances in machinery and linking this to a great flair for effective design.

Throughout the many volumes connected with the Expositions, the comment is frequently made that the French command of foreign markets relied more on the beauty of design than on any superiority of material. It was in this area that the dollmakers were so successful, making sure that their products stood quite apart from the very basic playthings made in Saxony.

Some idea of the importance that should be attached to the winning of an award in the mid 19th century can be gained from the fact that, at the 1844 event, out of some 3,960 exhibitors, 3,250 received marks of distinction. This showering of honours was a feature of almost all the nineteenth century exhibitions, as it was necessary to attract a large number of participants in order to make the display a success. Few manufacturers would have returned to the next event, with all the accompanying expenses, unless they were given some positive result in return for their effort: so anxious was the jury of the 1844 Exposition to bestow honours that a few were actually given to firms who did not finally exhibit. Seen in this context, the 1844 Honourable Mention to Jumeau and Belton, of Rue Salle-au-Compte, 14, is perhaps less remarkable.

According to the Jury, Jumeau and Belton displayed a fine collection of naked and dressed dolls, all being particularly well made.The company was already taking an interest in foreign markets and the Jury commented that their "large trade" included

Though many Parisiennes are attributed to Jumeau because of their pale blue elongated eyes, few are marked. This example carries an incised "J" at the base of the neck. Others are found without any head marking but with the "Medaille d'Or" body stamp. This example, dated by its original costume to around 1859 must be one of the earliest bisques made by the firm. She has a swivel neck, fixed eyes and the original finely plaited mohair wig. Her body is gusseted kid with padding under the leather to shape the breasts individually. The costume, with a small frill attached to the belt at waist level, is made of silk taffeta. The decorative chain-stitch machine edging was a clever utilisation of the sewing machine that was just coming into commercial use, again pointing to the progressiveness of the Jumeau establishment. Height 14 inches (35cms).
Author's collection. Photograph Acanthus.

items for export. Emile Jumeau claimed, in an American trademark application of 1888, that the company had used the "Bébé Jumeau" mark since 1840. Regrettably, this claim also seems to have originated in Emile's love of exaggeration, as even the term "bébé" was not popularly used at this time. Emile Jumeau was to make several other grand assertions regarding the family company, but contemporary accounts indicate that in the 1840's there was little, apart from the excellence of costume, to distinguish the work of Jumeau from any other. The fact that Pierre François was himself not without ambition is seen by the fact that he soon moved into his own factory at 18 Rue Mauconseil in the lst arrondisement, this new address first appearing in the 1848 Paris Directory, an address which the firm kept until 1867.

Natalis Rondot, in an 1848 report for the Paris Chamber of Commerce, mentions that there were some ninety makers of leather and carton dolls in the city at the time and that over eight hundred men and women were employed in the industry. He commented that great efforts had been made to introduce a more logical division of labour into the manufacturing process, so that prices could be lowered while at the same time the quality of the doll was improved. Much of the effort was put into an attempt to produce heads that had some chance of competing with those from Saxony, again emphasising native production. Rondot describes the various manufacturing processes centred around dolls including modistes, shoe, lace and even miniature flower-makers. He claimed that a Parisian working woman, giving her attention to a one franc doll, could create an exact reproduction of the fashion of the day. The Jumeau factory at this time formed only a small segment of this busy industry, but it was soon to attract the undivided attention of Natalis Rondot, who directed the Jury at the 1849 Paris Exposition. At this exhibition, the Jury commented that:

> "The leather doll is established in Paris with such superiority of work, taste and cheapness that it is sent to all parts of the world. It neither finds nor fears any kind of competition. This fact gives the display of M. Jumeau particular interest. His turnover is 120,000 francs, the majority of his products being exported and known as far away as China. He makes bodies of pink kid, of all sizes and qualities from 24 francs the gross naked, to fifty francs dressed. To make an attractive costume out of a few scraps of material is just a game to the Paris sempstresses, also we in fact insist less on exactitude and elegance of costume than on appearance and economy of manufacture".

The Jury goes on to compare two bills of costs. In one a carton or papier mâché head was obtained from Paris itself, the body was made of coloured sheepskin stretched over wood and it was sold for 15.6/10 centimes a piece. The second concerned a papier mâché head from Germany: with its carton body, arms of leather and hands of wood it cost, after dressing, 94 centimes. The first and much cheaper product was that of M. Jumeau. Rondot continued in his report:

> "M. Jumeau has, as one can see, almost achieved the limit of cheapness; he is at the same time one of those who has manufactured the best of Queens, Marquises and dolls of character. This artist-maker employs in normal times, sixty five women as cutters, makers of bodies, dressmakers and apprentices both in his own workshops and as outworkers. On piece work they earn from 1.25 to 2.25 francs a day. He obtains some of the heads from Germany and busts of wax from England; the character heads and carton bodies are made in Paris and the division of labour has already become such in this industry that the little stockings, shoes, hats, wigs, flowers etc. intended for dolls are items of distinct manufacturers.

The hauntingly mysterious face of the Jumeau Triste that holds continual allure for all who love antique dolls. The bébé has the characteristic fixed eyes and the applied, accentuated ears that are also a feature of this model. The original abundant mohair wig is still worn. The bisque head is incised "13" for the size, and the body bears the "Jumeau Medaille d'Or. Paris" stamp in blue. The heads of this doll, known in America as the "Long Face Jumeau" are never marked and the body should carry the Medaille d'Or mark. The bébé is ball jointed at the knees, elbows and shoulders. The original brown leather Jumeau marked shoes are worn with an extravagantly embroidered cream silk frock and a complex lace and silk bonnet. Height 28 inches (71cms).
Courtesy Emma Berry Collection. Photograph Acanthus.

Marked Jumeau heads with two rows of moulded teeth are extremely rare. This simple automata has fixed blue paperweight eyes, pierced ears and an open mouth. The bisque hands are very nicely moulded. The bébé appears from the centre of the silk flower holding out her hands. The head is marked "Déposé Tête Jumeau Bte S.G.D.G" with the size "3". Circa 1895. Height 9¾ inches (25cms). Courtesy Phillips, London.

M. Jumeau makes dolls' layettes and trousseaux from twelve francs the dozen to one hundred and fifty francs the piece. The twelve franc trousseaux are for nineteen centimetre dolls. They consist of nine items and the doll itself, but the popular piece at the moment is the trousseau layette at 4 francs 50 the box.This consists of a 24 centimetre doll and twenty six different pieces of costume. The workmanship is truly remarkable.The dressed doll is not only a toy but it often serves, dressed, as a model and pattern of our fashions, and it has become in these last years an indispensable accessory for every trip by our makers of novelties to the Americas and the Indies. M. Jumeau was, in 1844, an associate of M. Belton.This firm, at that time, received an Honourable Mention. The Central Jury appreciates the profitability of M. Jumeau's work and awards him a Bronze Medal".

The information included in Rondot's report is of great interest to collectors, as it establishes the fact that as late as 1849 porcelain heads were not used by the firm, though they were popular in Germany, and Jumeau's manufacture was still centred around the papier mâché shoulder heads which he obtained both from Paris and as imports from Germany. The very small sizes of the dolls mentioned is also curious, the 12 franc doll with "trousseau layette" measuring 24 centimetres (9.5 inches). Though the workmanship must have been of a high standard, it is difficult to understand how such miniature dolls could have provided dressmakers with sufficient detail and we have to conclude that it was the general effect of fashionable costume that was given rather than fine detail of construction. Of especial interest also are the wax busts of English origin which Jumeau costumed for the discerning French market. The detail contained in such Jury reports is rarely incorrect, as the individual firms were requested to reply to Questionnaires and several French Juries later in the century became quite irate because companies could not be bothered to supply the relevant information for their compilation. Probably at least part of Jumeau's success at exhibitions was due to the fact that the firm always provided an abundance of detail whereas the Germans, perhaps because of language difficulties, often sent exhibits without any background information at all.

Automata are only occasionally found with the original costume in such fine condition. This silk afternoon dress is very detailed. In its construction, perfect scale and elegant effect we see the characteristic flair of the Parisian dressmaker. The doll turns and nods her head as she "plays" with three movements. The 4¼ inch (10.8cms) cylinder has four airs. The original tune sheet is still in place on the back of the piano and lists these. The bisque head is incised "Déposé E 6 J", for Emile Jumeau. She has brown paperweight eyes and a closed mouth. Though Jumeau supplied the heads for such figures, the assembly was the work of the automata maker, who in this instance produced a very expensive parlour piece. Height 20 inches (50cms). Courtesy Sotheby's, New York.

At the Great Exhibition of 1851 in London, the Jury was disappointed because of the small number of toys on display, though there was a reasonable selection, including automata from Vienna and several effective mechanical items among the products of Bavaria. In dealing with the French exhibits, the report commented that, "Although France manufactures enormous quantities of toys of many kinds, only one description of them has been sent and that from a single exhibitor from Paris who exhibits dressed dolls only". Another French maker, Bontems, showed singing birds and A.J. Allix showed busts for hairdressers made of wax. The Jumeau award of a Prize Medal in the section is therefore hardly surprising. He had no real competition!
Pierre François, included in Class 29 and number 1282, France, won his Council Medal in fact not for the manufacture of the dolls themselves but because of the excellence of their costumes. The official report reads:

> "Prize Medal for dolls' dresses. The dolls on which the dresses are displayed present no point worthy of commendation but the dresses themselves are very beautiful productions. Not only are the outer robes accurate representations of the prevailing fashions in ladies' dress but the under garments are also in many cases complete facsimiles of those articles of wearing apparel. They might serve as excellent patterns for children to imitate and thus to acquire the use of the needle, with a knowledge of the arrangements of colours and material; in the latter respects they might indeed afford valuable instruction to adults".

English Juries were almost invariably concerned with the educational aspect of dolls and included them under the general educational heading, whereas in France they were part of Bimbeloterie, which included a whole variety of somewhat frivolous novelty type products.As at so many of the exhibitions, there was great controversy in London regarding the awarding of prizes, as there was no recorded account of the discussions of the Jury and the Press complained that the whole of the proceedings should have been kept open.
No attributable Jumeau dolls are known to have survived from this early period when the papier mâché heads were still being used. Pierre François was typical of many conservative French businessmen, who disliked change or dangerous and expensive speculation. Apparently he was always reluctant to introduce any new ideas and was, in typical bourgeois manner, mainly concerned that the security of his own family should not be endangered in any way. The eagerness of the firm to participate in foreign exhibitions and to work towards an increase in exports does reflect the willingness of the founder to increase the wealth of the firm and this was to form the basis upon which his successor could base his plans for more rapid expansion.
The "Travaux de la Commission Française", published by order of the Emperor, was not issued until 1855 and again comments rather regretfully that, out of some four hundred workshops in Paris making toys, only two exhibited in 1851 and "Only one exhibitor represented that curious industry of dolls which gives rise to a turnover of a million and a half francs". This section of the report was again written by Rondot who provided the additional information that "M. Jumeau, who is our most important maker of dolls, exhibited dressed dolls from 1 fr.25 centimes to 250 francs each. Trousseaux and layettes from one franc to two hundred and fifty for the box. The one franc box contained a nineteen centimetre doll and a trousseau of nine pieces".
Perhaps the most interesting part of this information that was contained in a footnote, is the fact that even in 1859 Pierre François Jumeau was already being described as the most important dollmaker in France, suggesting that though his methods were conservative, his business drive must have been great to have achieved such a position within ten years of the establishment of the firm.

An almond-eyed Jumeau dressed in an original French costume of pink and white ribbed silk, ribbon and lace and wearing a matching pink cotton bonnet trimmed with imitation fur and flowers.The unmarked head has a closed mouth, fixed brown eyes and pierced ears. The body is stamped in blue "Medaille d'Or" and is the early eight ball jointed type. The quality of the bisque in these portrait or almond-eyed Jumeaux is similar to that used on the early Parisiennes and gives them an ethereal effect. Height 15 inches (38cms). Author's collection. Photograph Acanthus

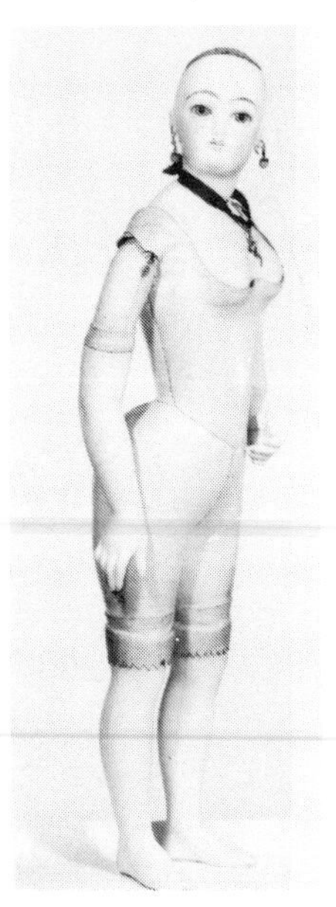

The undressed Parisienne has something of the amusing vulgarity of the late nineteenth century demi-mondaines. Although there is only one known example of this model with the leather body fully marked "Jumeau. Medaille d'Or Paris", any unmarked example of the same construction is attributed to the firm. This lady has a swivel neck and very well shaped breasts, for fashionable low necked dresses. The bisque arms also lend themselves well to costuming as they are bent realistically at the elbows. In the bisque feet are metal lined holes through which the doll could be fixed to its stand. Other examples are found with moulded shoes. Height 16 inches (40.8cms).
Courtesy Christie's, South Kensington.

Rondot, in a table concerning the cost of manufacturing the dolls gives an indication of the appearance of the Parisian doll of the time and mentions a bust of papier mâché from Germany, a body of carton, arms and hands of leather and legs of leather stuffed with sawdust, suggesting a type of doll similar to those termed "sabulas" by American collectors. It was apparently impossible for anyone to compete with Jumeau's work at this time, suggesting that the English Jury report dismissed his products a little too casually. Rondot continues:

> "It is in examining the trousseaux and the various elements of dolls' clothes that one better appreciates the degree of perfection to which this industry has been brought in Paris. There is not a single piece of their clothing which is not an exact model in form and fashion, so that the dolls are used in foreign countries as patterns for fashion". He added that the making of the porcelain dolls' heads in France had been affected by the fact that "The good porcelain painters would consider themselves denigrated by painting dolls' heads". The French Exhibition report lists Jumeau's Prize Medal: "Jumeau Dolls' trousseaux made with great taste and care. Dressed dolls which take French fashion to the foreigners and offer them at the same time as models for making clothes. A manufacture conducted with intelligence. Products which are cheap".

In 1855, at the Universal Exhibition in Paris, dolls were included in Class 25 and the official report commented that it was only in Germany and France that toys formed a significant part of the export trade. "Each country excels in the special types it has created, thus France distinguishes itself, as always, by the good taste and elegance of the dolls' costume".

Out of the thirty French exhibitors of toys, thirteen received awards. There were also forty seven foreign exhibitors, of whom twenty one were successful. They were divided as follows; one first class medal, seventeen second class and nineteen honourable mentions. Of the French makers it was Theroude (No.8926) who received the First Class Medal for his animal automata, while Pierre François took his place among the seven second class medals.

An early, very charming ball jointed Jumeau that can be dated by the original trained costume to the mid 1870's. The bisque head carries an incised rectangular mark, "Jumeau" within a border-type edging. Under this a size "1" and the letter "R". She has a closed mouth, pierced ears and eyes of the type more associated with Parisiennes made by the firm. The composition body is stamped "Jumeau Medaille d'Or Paris" in very bright blue. She has fixed wrists. The cream cotton dress is particularly fine with its ruched front panel and extremely long train. It is trimmed with both fine and knitted lace. The original hat is made of silk and trimmed with flowers. Height 11 inches (38cms).
Courtesy Abla Odell. Photograph Acanthus

"M.P. Fr. Jumeau. No.8918 Paris. M. Jumeau shows dolls which keep the cachet of elegance and good taste which have always distinguished the products of this firm, which makes not only expensive but ordinary dolls, with or without trousseau, all dressed with care, and whose prices allow for the significant sales he makes in France as well as for export".

The exports which a manufacturer achieved were of great importance to any French jury, aware of how desperately the Emperor was trying to increase national output, which had lagged behind that of their European competitors. In his address at the closing of the Exposition, he emphasised the fact that peace was absolutely necessary for the development of French industry and pleaded for other nations to stand beside France in resisting the Prussian threats of war. The doll industry made a small effort towards the innovations so necessary for the development of exports by displaying "Perfectly articulated dolls of a completely new new type. The French makers are showing some small articulated dolls known under the name of bébés". This term makes its first appearance at this event, and was used to describe dolls made by François Greffier, who worked from a small factory in Nantes and was rewarded with an Honourable Mention for an interesting new development. After this date, the term bébé was used by several makers and appears in the Paris Almanac of Commerce in 1859, though it did not come into general use until the late 1870's, when it was used to describe dolls that resembled children rather than adults.

The Exposition report provides the collector with some additional information regarding the Jumeau business, as a second class medal was also awarded "As a co-operator to Mlle. Delphine Floss, the senior worker at M. Jumeau's factory. Maker of dolls for five years, intelligent and adroit, she has contributed greatly to the success of the firm by the creation of new models and by the care which she brings to the execution of orders. Her conduct is exemplary. She maintains by her work her aged mother and three young children".

Regrettably, descriptive information concerning the dolls' heads is not provided, and we have to assume that this somewhat conservative company was still using the traditional carton heads, though the 1859 Paris Commercial Directory reveals that porcelain had become one of the materials utilized by the firm. The interest of Jumeau in the use of a comparatively new material was typical of the gradual awakening of the French toy trade at this period. A variety of inventions included a patent registered in 1855 by Blampoix for the insertion of glass eyes into porcelain heads, while Mlle. Calixte Huret had patented articulated dolls' bodies and was to invent a swivel neck in 1861. The early Jumeaux porcelain heads are a great source of controversy, the problem being made more difficult by the fact that the term for glazed porcelain and bisque is the same in French and we have to depend in this area upon contemporary reports, as no lady dolls with ceramic heads that can be positively attributed to the firm in this period have been discovered. As the dolls themselves are obviously unrecognisable, it seems most likely that German-made heads were used, but as there is no documentation, it is impossible to be certain, especially as it is quite possible that glazed, pink tinted French shoulder heads, similar to those used by Rohmer were utilised.

Standing by her contemporary doll's piano, still with its original candles, is an early portrait type Jumeau. She has fixed pale blue eyes and a closed mouth. The lips are painted in shades of pink with a slight gap between to suggest a slightly open mouth. The bisque is very pale and smooth. She has an eight ball jointed body in composition and wood with fixed wrists. The body is stamped at the base of the spine in mid blue, "Jumeau. Medaille d'Or. Paris". The head is incised with the size "8". She wears contemporary English underwear comprising drawers, several embroidered whitework petticoats and a feather stitched flannel petticoat with an original pale cream silk frock decorated with pintucking, embroidery and lace. The imitation fur stole is lined with silk and the head was given tiny glass animal eyes. She also possesses a good factory-made Jumeau winter coat in soft light brown wool with a red silk lined hood and a shaped full back. The coat shows a clever use of cutting and a decorative use of quite ordinary buttons. Height 18½ins. (47cms).
Authors collection. Photograph Acanthus.

Parisiennes

The French lady doll, termed a Parisienne at its time of manufacture, really came into its own in the 1860's. There was an air of intoxication in Paris at that time, with the dazzling fashion conscious Empress Eugénie showing the world the effectiveness of French dress at each public appearance. There was, in the streets of Paris, an almost universal desire for continual enjoyment and the Emperor, in his attempt to keep the affection of the people and create employment, encouraged a seemingly unending series of fêtes and festivals for which the streets and parks were especially decorated. As the native industry depended so heavily on the manufacture of charming, stylish, yet frivolous items, he saw the necessity of maintaining an extravagant court, with an abundance of display to provide the working class with work, as well as a continual spectacle. Each year there was a succession of pageants, the Carnival Procession of the Fat Ox, the Mid-Lent or Washerwoman's festival, the Gingerbread Fair at the Barriere du Trôn and the Fête Napoleon on August 15th. For each of these events, great schemes of decoration and illumination were put into effect and brought into Paris many people from the provinces, who were quite happy to spend their money on the charmingly contrived gifts offered in the small shops.

Parisian women, though perhaps inhabiting poorly furnished rooms with the damp paper peeling from the walls, threw themselves into this stream of enjoyment and paraded in the most daring and fashionable costumes. Every week some new design was popular and it became necessary for hats and collars to be adapted so that the woman, be she a marchioness or a dressmaker living in an attic, should be completely up to date.

For the French child, the grand occasion of the year was New Year's Day, the giving of presents in the 1860's achieving such proportions that in some homes the gifts were put on formal display. In order to allow even more vendors into the streets of Paris, the authorities permitted extra stalls to be erected for the sale of seasonal gifts, so that the great boulevards from the Madeleine to the Bastille became transformed into a double line of ravishingly alluring stands, where sweets and penny toys could be purchased alongside the well dressed dolls and Polichinelle. The French love of their own children became very evident at this time of year, and many stalls were forced to remain open all night in competition with the expensive toy shops, such as that of Giroux.

The falling birth rate in France, that so disturbed successive governments, was largely the result of the widespread use of contraception in advance of other European nations, middle class parents seeing no purpose in having to divide their wealth among a number of offspring. Even the peasants saw little pleasure in dispersing their life savings, so that in the countryside as well as the towns, a small family was normal. These favoured children were consequently showered with attention and given presents of much greater extravagance than children of comparable social standing in Britain or

It is unusual to find a Jumeau bébé dressed as a woman. In this instance, it is impossible to tell whether the costume might have come from the Jumeau workrooms or whether it was made after the purchase of the doll by its original owner, as the outfit is certainly contemporary and made for this figure. The design would suggest that the dress is of British origin, but it should be remembered that special dolls for particular events or for sale in a specific country were created by the firm. The fine, pale bisque head is incised "E.J" with the size "9". It also has the red decorator's tick mark. It has a closed mouth, pierced ears and the original auburn wig. The jointed body is marked "Medaille d'Or, Paris". The original underclothing that includes a practical flannel petticoat also suggests an English needlewoman. The dress and coat are dark Prussian blue wool with black velvet ribbon decoration. The bonnet is also black velvet. The shoes are marked "1878", which accords with the style of the costume. Height 22 inches (56cms).
Courtesy Sotheby's, London.
Sotheby's photograph. Acanthus arrangement.

I fondly truly love thee

Germany, who would have shared their parents' generosity with several brothers and sisters. There seems, in France, to have been some positive encouragement of a child's taste for expensive playthings and girls were expected to dress their dolls in the most up to date fashions. Under the mother's guidance, they would be taught to sew an elegant ensemble and create an eye catching accessory from appropriate scraps, thus learning the essential basic skill which would be needed as an adult, so that her own costumes should never look a week out of date. She was helped in the construction by patterns contained in special magazines, such as *"La Poupée Modèle"*, published between 1863 and 1869, that was intended to teach girls to look after their dolls just as their mothers looked after them. The magazine included notes on what the fashionable doll was wearing and, just in turning the pages, we realise what an exhausting occupation keeping up with fashion must have been at the time.

The magazine *"La Poupée"* included a number of patterns for dolls produced by Jumeau, such as a Russian bodice for an eighteen inch Parisienne, suggested in September 1863. This neat little garment was made of white silk trimmed with velvet ribbon with small bows at the neckline and wrists. The matching skirt showed the contemporary extravagance with material, as it measured thirty nine inches around the hem. In the next year, dolls dressed in a variety of costumes from shepherdesses to duchesses were offered, with frocks to be made of silk, tafetta, wool or the very expensive cashmere. Although Jumeau was to exhibit Parisiennes wearing cashmere shawls, it is interesting to note that shawls were only rarely worn by fashionable women in France, so that the figures were obviously directed more at foreign markets, especially America and Britain. A print dating to about 1863 shows two Jumeaux dolls with a spare jacket on a stand between them. It is impossible, because of style of drawing, to know whether the shoulder heads were of porcelain, papier mâché, or even wax. Both wore flower and feather decorated hats perched towards the front of the head, and the wigs were short and full of small curls. They represented girls rather than ladies, as one wears long drawers beneath her calf length skirt, which, of course, no woman of the period would have done. The other doll wore fashionable full sleeves under a bolero jacket, both outfits revealing an extensive use of braid and lace.

Unfortunately, the artists employed by magazines all concentrated upon making the dolls look as much like real people as possible, so that even Huret dolls, usually quite recognisable, appear in the line drawings like any child of the period. Some publications were used as an advertisement for a particular shop or maker, such as *"Gazette de la Poupée"*, which was promoted by Mlle. Huret, who organised soirées to which little girls could bring their beautifully costumed Parisiennes.

The 1867 Paris Exposition, the culmination of this period of reckless extravagance, put the city on show to the world, almost it seemed, as a feverish challenge to Prussia, with whom war seemed inevitable. In order to show foreign visitors that all was well, the whole of Paris was busy repairing drains and roads and decorating the shops. Among the more colourful visitors were the Prince and Princess of Wales, complete with footmen and Scottish pipers, though it was the spendthrift Americans who were most liked by the small tradespeople, whose business boomed between April and October as a result of the Exposition. Pierre François Jumeau, now established at the Rue D'Anjou-Marais, later the Rue Pastourelle, also sold many dolls to foreign visitors and several Parisiennes are reputed to have been purchased at this time; they must have made a very typically French souvenir.

At the Universal Exposition, dolls were included in Class 39, Group 4, as part of

Portrait Jumeaux were made for a fairly short period, but because of their great beauty they have long appealed to collectors. This example, with fixed blue glass eyes, has a softly painted closed mouth and a swivel neck in the bisque shoulder-plate. The gusseted body is made of white kid and has the "Medaille d'Or" mark. The head, as is usual on these portraits, is unmarked. She wears the original striped silk afternoon dress and has the original mohair wig. Her high heeled boots are particularly detailed and were probably originally some kind of sample footwear. Height 28 inches (71cms).
Courtesy Abla Odell. Photograph Acanthus

Bimbeloterie. The report on this section was written by M. Jules Delbruck who, after discussing the importance of toys in the development of children, regretted that the displays did not correspond with what might have been expected, and was struck with the disorder that seemed to reign in the showcases that were arranged without any concern for the child's age or needs. He felt that a toy should be useful, amusing, well made and cheap: instead he considered that the majority on display were motivated only by profitability and were constructed without any real logic and in pursuit of a dimly perceived fantasy.

Despite such basic criticisms, all agreed that the French display at the Exposition was brilliant, and constantly attracted great crowds, drawn by the roll of an electric drum. The best contemporary account was written by G. C. T. Bartley in November 1867, who again complained of the absence of cheap, ordinary playthings, and commented that by far the largest number of exhibitors in the section came from France, the majority from Paris itself:

> "The most important French toy is undoubtedly the doll and hence the visitor is not surprised to find that it forms the most numerous article in the cases round the attractive, though crowded and confined, French Toy Court. The manufacture of these puppets is carried on in Paris to a greater extent than in any city in the world; and as regards magnificence of attire and display of fashion, nothing can approach the French doll".

This English writer also disapproved of the magnificence of the Parisiennes with their over-elegant appearance and seems to long for the plainer London waxes but he does provide an invaluable description of the Jumeau factory at this relatively early period.

> "The manufacture of dolls is itself a very interesting process, though difficult to witness owing to the various portions being generally carried on in different places. M. Jumeau, at 8 Rue d'Anjou au Marais, is one of the largest makers in Paris. He employs many hands, the greater number being women, who are scattered in all parts of the city. The heads of his dolls are porcelain and most of the bodies of sheepskin, stuffed with sawdust, except the hands, in which iron fittings are inserted, to enable the fingers to move easily. The process of cutting the leather is peculiar, being done by hand, with an iron stamp set in box-wood. The stamps of course, vary in size and several different shapes are required for each doll.
>
> When the leather is cut the next process is to sew the parts together; this requires a regular apprenticeship, more particularly for the fingers which require great care; after the various parts are sewn up the body has to be stuffed and the limbs attached. The doll is then ready to receive its head; the manufacture of this part is totally distinct and is similar to that of ordinary porcelain. Cheap heads and shoulders are all in one piece, and their eyes are simply painted; while the superior descriptions are made separate from the neck and shoulders to enable the head to move on a sort of joint, and glass eyes are inserted into the sockets left for this purpose. The last thing is the hair and in this branch we much excel the French. At this manufactory, due to the heads being in china, the hair has to be put on as a wig and cannot be inserted so naturally as in a wax head. Human hair is rarely if ever used in Paris, the general material being mohair for the best and a sort of fur for the cheaper type of doll. The dolls when completed have to be dressed, and this process varies with the fashion, it would not do for a French doll to be behind hand in that respect. At M. Jumeau's establishment, the same style is not used for more than a month. All the dolls' clothes are made on the premises, where a roomful of young women is continually at work for these young fashionables.
>
> The dolls exhibited by M. Jumeau are not his best specimens nor can they be looked upon as giving a fair idea of his general style. Some of the smaller ones which are put in the less attractive parts of the stall are good. The three figures at the back of the case are got up in an elaborate and brilliant manner, representing ball costumes, but they are certainly not commendable as toys or suitable to the taste of English mothers for their children. The small boxes of dolls' clothes, or the dolls' trousseaux are good toys and from the variety exhibited and the very large numbers exhibited it is evident that they are very popular.

An early eight ball jointed bébé Jumeau with the pale face and more naturally sized eyes associated with the first of the articulated models. The head is incised with a faint size "3" but she is otherwise unmarked. She has a closed mouth and pierced ears with her fair mohair wig over the original cork pate. The eyes are outlined in black, more heavily than in the later bébés. She wears the original costume in pale green satin, giving a coat-like effect and being obviously Jumeau made. Height 10½ inches (27cms). Courtesy Sotheby's, London. Photograph Acanthus

M. Jumeau has not exhibited any of his mechanical dolls saying "Mama" and "Papa" and crying when laid down. These specimens are well worthy of a corner in the Exhibition. He also arranges the works and dresses a large number of walking dolls, the patent for which is American and for this reason he does not show them".

Bartley's account of his visit to the Jumeau factory makes it clear that the lady dolls produced in 1867 were sold with both fixed and swivel necks and that the cheaper grades of doll were given animal hair wigs. Human hair was apparently hardly ever used in Paris, which is why dolls of this period refurbished with modern real hair wigs look so curiously wrong. Though Parisiennes with painted eyes are now liked by collectors because of their comparative rarity, they were originally the cheapest type and presumably made in the greatest number. Comparatively few marked Jumeaux have survived from this period and I know to date of none of the painted eye type. Perhaps these, being unmarked, are erroneously attributed to Huret or Rohmer, as they are more typical of the marked examples of these makers' work. Obviously the collector should observe great caution in attributing any doll of this early period that is unmarked, as the similarities between individual makers are probably even greater than generally believed. The reference to the mechanical doll that cried when laid down is also tantalizing, as no authenticated examples of such dolls have survived. Perhaps, like the mechanical walking dolls, these were marketed rather than manufactured by Jumeau and were consequently unmarked. In either case, the description opens a further area of speculation. The American walking doll which Jumeau assembled and costumed was presumably the Autoperipatetikos, patented by Enoch Rice Morrison in 1862 and found with a variety of heads from the cheapest papier mâché to nicely modelled German white bisque. The quality of the costume also varies greatly, the best having the trained silk skirts of the 1860's while the cheapest were simply gathered cotton or dark wool. Again no examples are known that can be definitely attributed to Jumeau's costumiers, but the reference serves to indicate that the firm was quite happy to sell, if not to exhibit, dolls made by other people.

Dolls' costumes of the late 1860's were without parallel in the development of the French toy industry, as it was a period when a considerable amount of detailed work, such as the application of braids, sequins and beads, could still be carried out very cheaply by women who took real pride in their work, and, in typically French manner, were delighted, despite their own poverty, to see rich children carrying the dolls on whose costumes they had expended such patient work. The envy in a child's eyes as she looked at another's doll meant even more work for the sewing women. The very finest wools, silks and satins were used for the dresses, whose designs, as Bartley commented, were not repeated after one month. The firm of Jumeau was unusual, in that all this dressmaking was carried out on the premises, presumably still under the control of Mlle. Delphine Floss.

The only marks found on Jumeaux dolls of the late 1860's is an incised "J" at the back of the head. As the letter is incised at the base of the swivel neck it is sometimes difficult to differentiate between a "J" or a "1" so that great care has to be taken before making a positive attribution. The illustrated Parisienne in brown has well made clothes, but it is noticeable that they are without the spectacular extravagance of those marketed, for instance, by Simonne, and it was this effect, in combination with economy, that ensured prizes such as the Silver Medal gained at the 1867 Exposition. These Jumeau Parisiennes had to be effective, but also cheap enough to appeal to foreign merchants, and marked examples seem to accord with this, underwear, for instance, being usually

A richly costumed closed-mouth bébé with fixed blue eyes, pierced ears and unjointed wrists. She wears a cream silk and lace frock with a lace shawl. The bonnet is particularly ornate, and complements the doll's somewhat serious expression. The jointed body was divided at chest level for the insertion of the original pull-string voice box. The head is stamped in red "Déposé Tête Jumeau Breveté S.G.D.G 12". A decorator's mark, "XM" is painted in red. The body is stamped "Jumeau Medaille d'Or" in blue. Height 27 inches (69cms).
Courtesy Dorothy Brooke. Photograph Acanthus.

of a fairly economical type. In order to serve all facets of the market, a whole variety of costumes could be purchased, in addition to the basic doll, making it possible for the house of Jumeau to supply foreign buyers of all types.
The early Jumeau Parisiennes have completely unmarked hand-sewn leather bodies with inserted gussets that allowed some movement at thigh and knee. Though the wire that formed the armature was originally useful both for strength and to allow the limbs and fingers to be variously positioned, it has often failed to stand the test of time and the leather is sometimes stained with rust, especially in the area of the fingers. In some instances, the wire actually protrudes and has damaged the surface leather. Later, more adventurous constructions were attempted but in the 1860's Pierre Jumeau seemed to be concentrating upon a well made and economical product rather than on innovatory ideas. The bisque heads of these dolls exhibit the characteristics that were to be a feature of Jumeau lady dolls until the 20th century with fairly large, elongated eyes, often of a sharp, pale blue and gently defined features. As with dolls made by every maker, the quality of the bisque and the colouring varies, so that each Parisienne has to be judged on its individual merit.

The years of change

The 1867 Paris Exposition marked the end of a period when France seemed to have spent all her energy on the creation of frivolous items of sumptuous effect such as the lady dolls with their huge trousseaux. Even during the exhibition, while the Emperor and court attempted to keep up appearances, the constant worry regarding the Prussian army was always near, casting a cloud and yet inspiring the pleasure seeking visitors to greater excesses. Since the time of Napoleon, and despite eventual defeat, the French people had considered their army the finest in the world and young men in military uniforms were a feature of every-day life. It seemed impossible that such splendid soldiers could be vanquished. When the Franco-Prussian war came, the horror of defeat was even greater, as it followed the years full of fêtes and pleasure. During the conflict, industry, hampered by lack of funds and the poverty of its traditional markets, was almost at a standstill, while business transactions became extremely difficult as many banks closed their safes and refused to return deposits. During the siege of Paris, some 80,000 citizens were killed and many lost their homes and life savings but more damaging was the loss of national pride and the humiliation of the army.
Though the losses of the war were great, more hurtful was the reparation demanded by Germany in gold and silver, reparation which at first it seemed quite impossible to pay. Strangely, when all seemed darkest, the French national character asserted itself and the whole country plunged into a great effort to re-establish industry and free the country from its onerous debt. In fact the recovery of France after the war was so rapid that it caused the Prussians to become more wary of its defeated neighbour. Business, literature and the arts all took on new vigour and seemed, because of the war, to have developed an awareness of the real needs of the late nineteenth century. Much of the traditional caution regarding the preservation of family wealth became tempered after

The wood and composition Jumeau body is sometimes found shaped in an adult form; the main difference being in the modelling of the torso, which is waisted and has suggested fullness over the breast section. This pale example wears its original animal hair wig with a cork pate. It has fixed brown eyes and pierced ears. The closed mouth is softly represented in pink. She wears an attractive original costume of a knee-length coat over a floor-length pleated skirt, both made of pale blue/grey grosgrain silk. The coat edge is decorated with lace and braid edging. A fashionable large burgundy velvet hat sets off the ensemble. These lady Jumeaux are always popular with collectors, as comparatively few were made. The head is stamped in red "Déposé Tête Jumeau Breveté. S.G.D.G". There are artist's tick marks and the size "10". The body carries the oval cream "Bébé Jumeau Diplôme d'Honneur" label. Height 24 inches (61cms).
Courtesy Abla Odell. Photograph Acanthus.

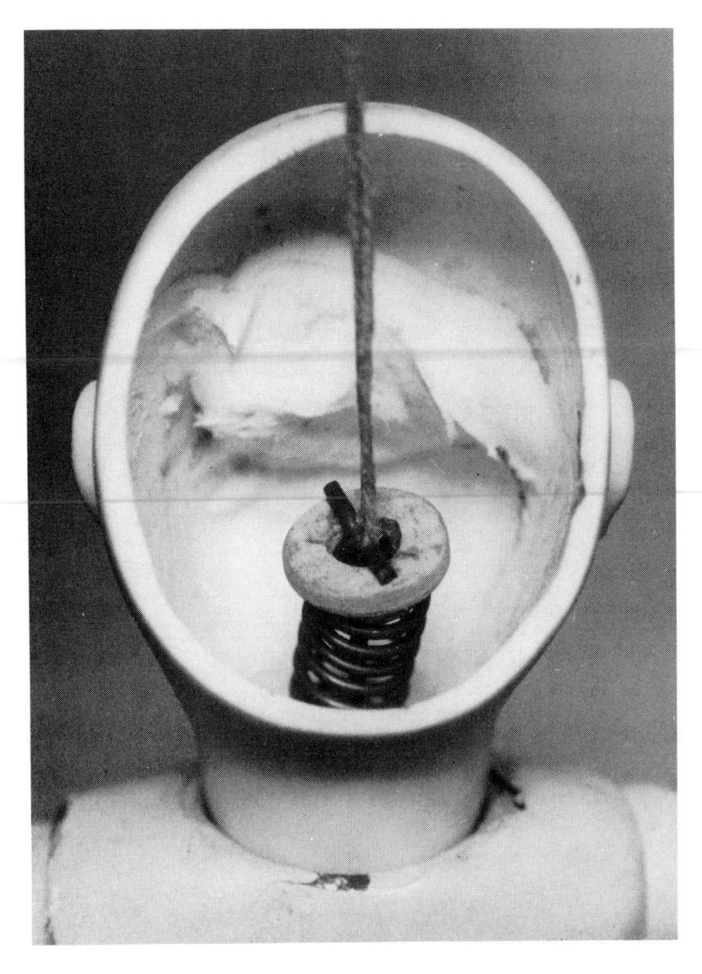

Interior of a bisque Jumeau head of the "Diplôme d'Honneur" period. The tension of the stringing was maintained by the use of the heavy metal spring. This example still has the original cord that was used during the stringing process.
Author's Collection.

this defeat, which in itself helped the more adventurous companies. Paris, despite its suffering, soon re-asserted itself as the fashionable centre of Europe and all the great events began to be staged, a little different, but just as elegant, so that the races, for instance, took place on the Auteuil course instead of La Marché, but were just as colourful and grand. Light poured into the Louvre gardens because of the destruction of the Tuileries and in other parts of the city people commented on the new, bright light that seemed to illumine everything because of the destruction of so many buildings. Such was the buoyancy of the French character that, instead of weeping at this destruction they decided that ruins were picturesque and strolled among them with their children, all fashionably dressed, with the women wearing that new creation, the bustle. Fashion was perhaps a little less extravagant, but prosperity returned together with the speculation, the gambling, and the daring young men who rode around the city on their velocipedes. Though the effect of the war had been considerable, the doll industry, because of the cessation of German imports, had gained in strength and there was considerable activity in this trade, as in all commercial life.

A swansdown and lace bonnet complements this closed mouth bébé, with its unusually wide-eyed appearance. The bisque socket head is stamped "Déposé Tête Jumeau Bte S.G.D.G 11". Inside the head is the decorator's brush mark, the letter "W" in pink. The S.G.D.G. mark means that the doll was sold with a patent but without a guarantee by the French government. This meant that if the patent was infringed the individual maker would have to deal with any legal action himself. The body, with jointed wrists, carries the oval "Jumeau Medaille d'Or" label in cream. She has a closed mouth, blue fixed eyes and pierced ears. A contemporary whitework embroidered frock is worn. Height 25 inches (63.5cms).
Courtesy Emma Berry Collection. Photograph Acanthus.

Several changes took place in the Jumeau establishment during this period and, probably encouraged by the lack of German dolls, a larger factory was set up in Montreuil in 1873. Here was made everything necessary for the retailing of a doll, including the boxes in which they were to be packed. At this time it was expected that Pierre François' eldest son, Georges, would eventually take over the factory. Georges was of a much more adventurous character than his father and often tried to persuade him to abandon old customs, and did in fact manage to introduce some small reforms. His death, when he was within sight of taking over the firm, apparently devastated his father, who decided to retire. At this time the company had an annual turnover of about 150,000 francs. Emile Jumeau, the second son, originally intended to be an architect and is often thought to have taken control at very short notice but in fact seems to have been involved for some time in the family business, as he was mentioned for co-operation in the manufacture of dolls, in an award at the Vienna Exhibition of 1873. The dolls shown at the Vienna Exhibition, where a Medal of Progress and a Gold Medal were won, attracted the enthusiasm of the jury, in whose report the heads were described as of "Enamelled porcelain of the greatest perfection". Jumeau had surpassed in beauty the products usually brought from Saxony, The Jumeau display was not only impressive but the merchants were also surprised by his low prices. The medal for co-operation for the production of dolls was shared jointly by Emile Jumeau, Madame Blanche Pannier (for her fine hats), Eliza Cadet and Oscar Rinder.

Just a few years later another gold medal was awarded to the firm at the Philadelphia Exhibition for "A fine collection, dressed in the most fashionable style". It was in the same year, 1876, that Emile Jumeau took over the family company on Georges' death, though there is some confusion over this date, and Emile later claimed that he took over in 1875. Under his much more ambitious leadership the company seemed to come to life and every technique and advertising ploy was used to make the public both in Europe and America aware that the important name in the doll industry was Jumeau.

The 1878 Paris City Directory describes Emile Jumeau as "A manufacturer of kid dolls and jointed wooden dolls", and surviving examples in original costume suggest that it was still dolls of the traditional lady type that were made, though there had been some advance in the design of bodies, and the more complex versions of jointed wood, usually painted cream, were also in use. This type of lady doll meant that the figure could be positioned in various ways and avoided the stiff look of the sawdust-filled leather. Emile claimed to have been experimenting during these years with various types of body construction, as the dollmakers were obviously aware of the limitations of the lady dolls, and were conscious that children were beginning to demand change more rapidly, and that their doting parents were eager to supply them with dolls of even greater beauty.

Any nation embraces its children with fervour after conflict, and the attention and money showered on the children of Paris actually offended many visitors. Albert Rhodes, writing in the mid 1870's, described the Champs Elysées, where "The favoured children of fortune were riding behind gaily decorated goats and playing in the gravel, mounted on revolving horses of wood and spearing at the pendant rings, like knights in a tournament, while their nurses stood by and clapped their hands whenever they were successful". Other children played in the swings or boats, others enjoyed make-believe games and a larger group watched a puppet play. Around the booth a cord was stretched to mark the line between those who had paid and the children who had crept up to watch. On benches sat well dressed children with their mothers watching fashionable

A most beautiful portrait type bébé, the bisque head impressed "E.J." for Emile Jumeau and with the size "9". The body is of the eight ball jointed type with the early fixed wrists. The torso is stamped in blue "Jumeau Medaille d'Or". Hands are often a weak feature of even the finest dolls, but here they are well shaped. The doll has applied ears, brown fixed eyes and a closed mouth. She wears the original mohair wig and a contemporary smocked silk frock. In the trunk is a group of German all-bisque baby dolls. Height 23 inches (58.5cms).
Courtesy Emma Berry Collection. Photograph Acanthus.

young Paris at play. "Here the cup of joy runs over. Riding behind goats and wooden horses, ascending and descending in a circular swing". Wherever he looked, all was perfection and light.

Such adored children needed the most beautiful dolls to complete the perfection of their lives and Emile Jumeau was ready to supply them. The factory in Montreuil, first occupied in 1873, was situated on a convenient long straight road into Paris, which meant that the dolls could be in the main showroom within hours of leaving a dressmaker's hands. The factory, rebuilt by Emile, was considered very progressive and looked more like a school or municipal building, as the main blocks enclosed a pleasant courtyard with a fountain. This smart, new, gas-lit factory seemed to demand a different product and by 1878 Emile had perfected his new, somewhat more substantial type of doll's body which, in combination with a child-like head, became known as a bébé. The Paris City Directory for the first time in 1879, mentions "unbreakable bébé dolls" in Jumeau's listing and describes them as "Unique models". It seems probable that these were first shown at the 1878 Paris Universal Exposition, at which a Gold Medal was awarded to the firm. In the official report on dolls by M. Rossolin, the comment is again made that the industry was essentially Parisian and that dolls were one of the sections of the toy trade that had achieved most progress since 1867. Included in the class were:

> "Dolls and naked bébés in kid, fabric, wood, moulded carton, bisque, porcelain, wax etc. Dolls and dressed bébés of all kinds, doll accessories, wigs and hats, shoes, jewellery etc. Almost all kinds of dolls were shown at the Exposition, however we were able to notice a few gaps, such as the very common dolls in moulded carton and the old poupard. One must add that in the making of dolls, good French taste has been used to reproduce as exactly as possible both woman and child, both from the point of view of the modelling of the body, from the expression of the face and from the point of view of movement. The bébé, which better satisfies the taste of children, has tended to supplant the poupée (lady doll. Trans) and it does not seem unlikely to us that if the former were produced as cheaply as the latter sales would increase considerably".

Though Jumeau was concentrating mainly on bébés, one of the most beautiful Parisiennes dates from the Medaille d'Or period. It was by far the most complex figure attempted by the company, with a swivel neck and moulded breasts. The feet are also unusual in that they are made in the correct position for wearing high heeled shoes. The well shaped bisque arms and legs contrast effectively with the complete lack of anatomical detail seen in the leather bodied dolls. The central section of the body, that sometimes carried the Medaille d'Or mark, is that of a typical Parisienne, but the complete doll gives more the impression of a figurine rather than a play doll. One example still retained its original eighteenth century style fancy dress costume, which added to the appearance of a pretty statuette. Despite the effectiveness of such dolls, they were the last to show any positive advance in the design of Parisiennes, as Emile Jumeau's attention was directed towards the perfecting of the much more progressive bébés.

Jumeau claimed to have studied the design of a successful body for bébés for many years before he finally produced a figure capable of mass production, and the City Directory of 1879 added to his listing as a manufacturer of "kid dolls and jointed

A Jumeau Triste in such completely original costume is only rarely found. This outfit is created from printed muslin decorated with cotton lace and trimmed with perfectly co-ordinated magenta ribbon. The cape collar, in-filled with lace over muslin is simple but effective. The whole of the costume is machine sewn. The cream silk bonnet is trimmed with gauze and decorated with magenta flowers that perfectly match the ribbons on the frock. She wears lace mittens that match her original lace socks. The cream leather shoes have metal buckles and are marked on the sole "Société Marché Simonet 95 rue de Sevres". The doll has fixed brown eyes and applied ears. She wears the original mohair wig. The ball jointed body is stamped "Jumeau Medaille d'Or". As in all Jumeaux Tristes, the head is unmarked but carries an incised size number "11". Height 25 inches (63cms). Courtesy Abla Odell. Photograph Acanthus.

To the surprise of collectors, fine Jumeaux are still occasionally found in attics. The owners of this example had no idea of the value until it appeared in the sale-room. This Jumeau Triste, in its contemporary red silk frock, is very appealing, as it is so completely original. The head is marked with its size number "16" and the body stamped "Jumeau Medaille d'Or. Paris". Height 33 inches (84cms). Courtesy Christie's, South Kensington.

wooden dolls" the phrase "Unbreakable bébé dolls" that were unique models, and it was obviously in this year that the Jumeau promotional machine really concentrated upon the bébés that had been unveiled at the Exposition. A quantity of the old style Parisiennes continued to be manufactured until the end of the century, but it was the new, articulated bébés that most fascinated Emile. Although the Parisiennes continued to be costumed with great care and in the most fashionable clothes, because of the continuing public demand, the Jumeau company considered them very lightly, as this progressive house saw the articulated child-like dolls as the great advance, upon which their world wide fame would depend.

Of all the various Jumeau products, it is these early ball-jointed examples that are most liked by collectors, as their importance as a dramatic stage in the evolution of the design of dolls is generally recognised. The early bébés have fixed wrists and separate ball joints at shoulder, elbow, thigh and knee, which allowed for a great deal of movement. As the legs on these child-like dolls were much more substantial, it was possible for the figure to stand unaided, which made them especially attractive. Before the advent of the bébé, it was almost a convention that the feet of dolls should be unnaturally small, which made it extremely difficult to balance the figure sufficiently well. The very thickness of the new bodies made it possible for quite complex methods of articulation to be concealed within, whereas the makers of articulated wooden bodies had been forced to rely on precise jointing that remained visible. When I first began collecting dolls, there was a generally held belief among many established enthusiasts that some of these early ball-jointed bébés were strung with cord or string rather than elastic. I can only comment that, despite handling a number of dolls of this type in original condition, I have never discovered an instance where cord was used and its utilization would, in fact, have seemed a negation of the design which was based on a series of hooks and springs that put the figure under continual pressure. It seems probable that some of the early collectors might have found examples with the cord still attached to the spring inside the head, cord that was used as part of the pressure stringing process but having no direct connection with the articulation. Reluctant to investigate further, this thin cord was perhaps thought to be integral to the

This completely assured child was made by Emile Jumeau. She has fixed brown eyes, a closed mouth and pierced ears. The original mohair wig is worn. Her original green velvet costume with its matching hat trimmed with ribbed silk is very attractive. The bisque head is incised "Déposé E 6 J". The effect is increased by the original shoes and small contemporary purse. Height 15½ inches (39cms).
Courtesy Abla Odell. Photograph Acanthus

construction. It is only possible to speculate on the reasoning, but it does seem necessary to dispense at this point with an old tradition that can have no basis in fact.
The bébé bodies, made from very good quality composition, were remarkably strong and their only weak point seems to be the very thick layers of paint which, over the years, have sometimes cracked, causing quite large sections to break away, revealing the body of the composition. One rarely finds the trunk of a Jumeau body crushed, as is quite often the case with cheaper German dolls, as rigidity was maintained by the fixing of a transverse wooden bar, which was also used as an anchorage point for the series of hooks, springs and wire clips which held the figure together. Despite the skill exhibited in the manufacture of the bodies, it is the fine quality of the heads that first attracts, as the bisque was still of that texture and almost transparent quality associated with the better Parisiennes. The healthy, pink-cheeked children of the twentieth century were undreamed of at that time, and all have the wistful pallor of the mid-nineteenth century child, still protected by bonnets and parasols from too much exposure to the sun. In profile, some of the early heads are hardly attractive, as they have little definition of feature, and there is hardly any suggestion of a chin, but seen in full face they are haunting and wistful with large eyes, delicately painted but often heavy brows, and pale lips. The public obviously found these dolls extremely attractive, and production rose from a value of 150,000 francs in 1875 to more than a million in 1884. Pierre François Jumeau was apparently amazed at the changes in the workshops, which Emile claimed to run in the intelligent manner originally planned by Georges.
The 1882 Paris City Directory states that in the previous year 85,000 dolls were sold and that sizes nine to sixteen had paperweight eyes and applied ears, though it should be noted that in fact smaller sizes are also found with the more time-consuming, separately modelled ears. All were sold with a necklace of pearls, which was useful in concealing the neck joint, and a comb in their hair, but more expensive dolls were costumed and it was upon the amount of work and the cost of materials associated with this that the eventual cost of the toy was largely dependent. Collectors of today, concerned with the basic doll, almost to the exclusion of costume, are often surprised to discover that the object they prize so much was, in effect, just the shape upon which the Parisian costumiers could display their skill, and which would make the dolls stand out in any toy shop whether in Paris or New Orleans.

The Notice

In 1885, for the first time, the Paris Directory commented that all Jumeau dolls carried the firm's name and also added that the bébés had been marketed since 1879, one hundred and ten thousand being sold in 1883. With his great enthusiasm for publicity and a willingness to make use of every award gained in order to make both himself and the product of which he was so proud even more desirable to the public, Emile was succeeding in making the name of Jumeau familiar to all toy sellers, and, perhaps even more importantly, to the children for whom they were ordering. The 1880's were the summer of the Jumeau factory, full of new ideas, enthusiasm and rapid progress that encouraged the firm to publish booklets such as the "Notice on the making of Bébé Jumeau", printed by J. Cusset and issued in 1885. The tone of the "Notice" makes it

An unmarked bébé with a pensive appearance. On the back of the head are the decorator's tick marks (check marks in America), accepted as a Jumeau mark. On the jointed body is also the Jumeau oval cream label. This good quality bébé has fixed eyes and well moulded pierced ears. It still has the original cork pate and an old wig of real hair with a hairslide of the type often found on Jumeaux in original condition. A broderie anglaise pinafore is worn, with a whitework frock and a cream silk bonnet. The wrists are of the jointed type. Height 26 inches (66cms).
Courtesy Emma Berry Collection. Photograph Acanthus.

fairly obvious that either Emile himself or a most sycophantic employee was the author, though the title page gives no positive indication and reads, "Notice of the manufacture of Bébé Jumeau in the factory of Montreuil sous Bois. Followed by a history of the firm of Jumeau from its foundation".

The "Notice" took the form of a conducted tour of the factory and the author compares the little girls attending to their doll's toilette with the behaviour of mothers who see that nothing is too good for their own child. After all, the greatest pleasure anyone can give a little girl is a doll and of all dollmakers, the firm of Jumeau, universally known and with a world-wide reputation provides children with most delight. The enthusiastic writer, encouraged by his thorough knowledge of the factory, then embarks on the tour, by entering a large hall in which vats of Kaolin, a constituent of porcelain, were stored. These contained enough material to manufacture thousands of heads and it was here that the kaolin was allowed to macerate for long periods, as this improved the quality of the porcelain. When the paste was needed for the casting of the head it was taken from the water and worked in the hands to even the consistency and make sure that there were no air bubbles before it was rolled into a suitable thickness for the size of the head. The paste was then pressed into two part plaster moulds that could be used only fifty times before definition was lost: the larger heads, requiring the manipulation of thicker paste, were made by men, while the more numerous smaller sizes were, like so many other sections of the factory, the province of women.

After their removal from the moulds, the heads were allowed to dry out a little before the eye sockets were cut, a most delicate process, as it was neccessary to chamfer the edges so that the eye would fit perfectly and realistically. At this stage, the ears were applied to the larger heads, those of the smaller versions having been shaped in the mould. Presumably the ears were applied to the smaller sizes when a particularly good effect was required or perhaps for dolls made to the special order of a retailer. The heads were then placed in saggars which held as many as three dozen heads that were placed inside one another, according to size, for economy of kiln space, great care being taken of course that they should not touch. Such an arrangement of the kiln indicates how well controlled was the complete process, for with any real degree of risk the heads would have been separated more completely. Obviously kiln losses were minimal or the heads would certainly not have been put at further risk by being stacked inside one another, where any air bubble or small imperfection in the paste could have caused an expensive explosion. Some kiln losses were of course inevitable and the "Notice" comments on the amount of waste that lay around the factory in baskets resulting both from kiln damage and from faulty decoration.

The heads were fired for twenty seven hours in wood-burning kilns and, after cooling, were transported in huge baskets to a workroom for polishing. This very tiring process was again carried out by women, who placed the smooth heads on shelves in front of them. Soon they were again collected and packed into baskets to be taken to the two studios for painting. In the first, the undercoating of the head was applied then they were passed to more skilled men and women for the final coating of flesh colour, made from what are described as mineral colours, used because they were finer and apparently kept their colour longer, though this would again seem an exaggerated claim. Two coats of pink colour were first applied all over the neck and head and they

Few mulatto Jumeaux have survived in such wonderfully original condition. Too often, the factory clothes have been removed and replaced by conventional childlike creations. Here we see the original Jumeau metal bangles and beaded necklace in combination with a rich yellow silk gauze-trimmed skirt and a carefully pleated bodice that was sewn in position. She wears a simple petticoat and short drawers in light brown. That Jumeau did not intend the child to re-dress the doll is also seen in the way the bangles are fixed. The doll has a closed mouth, fixed brown eyes and a composition and wood lady-type body with the accentuated waist. The body is stamped on the back, "Jumeau Medaille d'Or. Paris", the head being stamped in purple "Déposé Tête Jumeau Breveté SGDG. 8". Height 18½ inches (46cms).
Courtesy Abla Odell. Photograph Acanthus.

were then dried in special chambers before the application of the cheeks, lips, eyebrows and eyelashes and the red nostrils. The greatest care was needed during this operation, as the mere slip of a finger could mark the head and it would have to be restarted. Once more the heads were fired, but this time arranged alongside one another for greater safety, and at a lower temperature. Seven hours sufficed for this final process. The "Notice" boasts that all other dolls were dethroned by the Bébé Jumeau and were forced to remain unsold for months in shop windows, faded and ugly, until the despairing seller threw them in the attic. "The reign of the German doll is over, Bébé Jumeau has succeeded. It will surely be a long reign". Despite the obvious popularity of the Bébés, whose manufacture involved the skill of at least thirty people apart from the dressmakers, part of the factory was still engaged in the manufacture of Parisiennes. These were, according to the "Notice", less beautiful than the bébés but, as a considerable number were still in demand, their manufacture would continue. No other firm, claimed Emile, could compete with his factory in workmanship even of the traditional lady dolls which, with neatly made leather bodies and swivel heads, were packed in boxes wearing earrings, a necklace, comb and a lawn chemise, to await their final up-to-date costumes.

Even in nineteenth century France, profit was obviously a word with undesirable connotations, so the writer comments that M. Jumeau was surely unable to make very much out of his dolls, as the cost of raw materials and labour was so high! In fact Emile Jumeau was very sensibly profit conscious, and handled all the export side of the business himself, dispensing with the cost of a middle-man, which made the dolls cheaper and better able to compete with the German products.

The manufacture of dolls' eyes, at the time of the "Notice" in 1885, was at the Rue Fontaine-au-Roi, but it was soon to be moved to the main factory at Montreuil. The women who made these very fine quality eyes served a five year apprenticeship and twenty five were at work in the studio. The philanthropic activity of Emile is much emphasised by the writer and we learn that he employed as apprentices only orphaned or abandoned girls. "The good works of M. Jumeau cannot be counted and each day he is congratulated by his many friends and encouraged by the League of Patriots, of which Jumeau has been a member since its inception".

In the workroom, the girls sat before gas-burners holding a glass tube about twenty five centimetres long in one hand and in the other a rod of black glass, which, by turning on the end of a glass tube, made the pupil. A stick of blue glass was then melted in a circle around this to form the iris, and, with a very fine piece of white glass, the small lines in the iris were made, lines which produced the so-called "human eye". The threads of glass had to be twisted as in a paper weight during this process so that many small facets were created. Though the writer made much of this process and almost suggested that it was especially created by Jumeau, these were techniques which had long been in use in the glassworks, especially for the making of artificial eyes.

After the basic work of setting the pupil and iris into the white of the eye was completed, clear crystal was then heated and dropped over the eye, a process known by Jumeau as crystallization. This gave the eye added brilliance and also acted, as in a paperweight, as a means of magnification. The eyes were then left to cool, but apparently given an additional firing to complete the process before being packed in boxes according to size. It is in detail, such as the construction of the eyes, that Jumeau was able to surpass his German competitors, who used the cheaper method of applying the glass for the iris and the pupil to the surface, consequently losing the depth that is a

Jumeau faces continue to surprise, no matter how many are examined. This beautiful bébé has fixed blue eyes, a closed mouth and well modelled applied ears. The head is stamped in red "Déposé Tête Jumeau Breveté S.G.D.G 13". She has a jointed body with the oval "Bébé Jumeau Diplôme d'Honneur" label and wears a contemporary child's cotton smock with very bold embroidery. Various versions of this design are found with the flowers in different colours. It is edged with coarse lace. Height 28 inches (71cms). Courtesy Rachel Karslake. Photograph Acanthus.

feature of the best French eyes. The "Notice" comments that Jumeau would shortly be introducing the "half spherical eye" with the pupil and iris made in the same way but with the basic white enamel round instead of oval, so that the bébé would be able to raise and lower its eyes.

The fixing of the eyes into the head was another detailed and skilled process, as it was initially held in place with a layer of wax and subsequently held permanently with plaster. Dolls were already in the process of design for 1886 with eyelids that were raised and lowered by means of a spring, and in order to facilitate this movement, the eyes were to be set deeper into the sockets to leave room for the moveable eyelids, which were to be fixed with small wires. From a purely aesthetic point of view, this advance towards sleeping eyes was a pity, as no eye that has to move can ever be fitted as perfectly as the fixed paperweight type in its chamfered socket, though the child customers probably welcomed the novelty.

At the Montreuil factory, it was only the apprentices who could be seen at work on the manufacture of bodies, as the majority were made at home by women who had learned their trade in the factory and were thus enabled to make a good living for themselves while looking after their families. The substantial bébé bodies were made of thick, grey paper soaked in paste and made of layers according to the strength needed for that part of the body. The parts were cast in halves in cast iron moulds, and then glued together and dried for twenty four to forty eight hours, depending on the season. The cups for the attaching of the elastic were then fixed, cups which could be made of metal, wood or cardboard. The parts were subsequently sent to the paint shop, a room where everything seemed to take place in the white haze of the ground colour. The various parts were then put to dry on sticks and later sanded smooth. In another room, five coats of pink paint were applied, then a final coat of varnish. These paint shops occupied a complete wing of the factory, which gives some idea of the scale of the operation. The assembly shop was on the ground floor and there, more than fifty people worked, stringing the dolls with elastic so they could be made to stand up quite straight or even put into poses such as saying their prayers. The head was then fixed in place and a cork pate applied so that the wig could be artistically arranged and made to curl prettily on the forehead and neck. When dolls with original mohair wigs are found, it is usually seen that the wig was in fact fixed to the cork with a few nails and then arranged and glued precisely in position. Mohair in a variety of realistic shades was used, from an attractive soft auburn through all the shades of fair to gentle browns and occasionally black, the latter used in combination with brown eyes. In the early types, the mohair was hand-knotted, creating a very light effect that suited the early type of bisque particularly well.

The "Notice" informs us that some long wigs were used with the hair cut into a fringe at the front. All the hairdressing was again a woman's speciality and the process was completed by tying a length of ribbon in the hair and adding a bead necklace. The dolls, at this stage, were costumed only in an embroidered, lace-trimmed chemise and stored away in boxes until October, when they were shipped abroad for the Christmas market. Some of the dolls made at Montreuil in 1885 had a speaking mechanism of a fairly simple type which, although it made them more expensive, still sold some 20,000 each year.

The completed, costumed dolls were displayed in the showrooms on the Rue Pastourelle, where both adults and children could be seen gazing for hours at the finely

The charming French wash-stand, complete with its original china toilet ware, dates to the same period as the dolls. The bébé in the cream bonnet has an open mouth, fixed blue eyes and pierced ears. The head is stamped "Tête Jumeau" in red and incised with the size "9". She has a jointed body with the oval cream label "Bébé Jumeau Diplôme d'Honneur". Height 21 inches (53cms). The closed-mouth bébé wearing the large brimmed bonnet has fixed blue eyes and a fully jointed body. She wears the original green shoes. The head is stamped in red "Déposé Tête Jumeau" with red tick marks and an incised size "8": Height 18 inches (46cms).
Courtesy Denise St Clair. Photograph Acanthus.

dressed figures. Inside the shop, there were showcases that were divided into scenes, such as a ballroom and a dining-room, and in these settings the dolls, in a variety of costumes, were displayed. They were not only costumed as children but included a Watteau shepherdess, soubrettes and a Marquis of the Louis XV period. So exquisitely costumed were the child dolls, however, that they seemed to forecast fashion itself. It was over this most important area that Madame Jumeau reigned with "Her impeccable taste, her perfect elegance, her love of beauty". She met all her workers each morning to see their latest designs for hats and clothes as "Many millionaires' children do not have more beautiful clothes than Bébé Jumeau, with colours that are artistically blended, the silks and satins, laces and ribbons, all carefully selected to form an artistic whole". Some of these frocks took eight days to complete, while a few of the ball gowns took even longer. The materials came from the factories at Lyon and all the laces and trimmings were also bought direct from the makers. Most of the sempstresses worked at home, but a few were employed at the factory itself. The needlewomen created fashionable models and then offered them to Madame, who ordered the requisite number so that more than three hundred new designs were used each year. For foreign customers, the dolls were sometimes especially dressed in national costumes, and Jumeau was in the process of preparing work for a special exhibition in New Orleans based on the theme of the Great Mogul, so that a variety of accessories, including cages and baskets of snakes, were in the process of construction in addition to the dolls.

The "Notice" boasts that many firms attempted to copy these splendid Jumeau outfits but their superiority was so great that the imitations were immediately apparent. "Besides, on the back of each bébé there is the full name". Various claims were made regarding the marking of the dolls, but it is often unclear whether some kind of permanent mark, such as those incised on the heads or merely a ribbon or label is indicated. As the more insubstantial markings were obviously soon lost, the firm's claims do not give the collector much help, though they serve to indicate the complete seriousness with which Emile regarded his product. In the "Notice", a hint is given that he was aware that other people found his devotion to his doll factory somewhat droll, but, as he provided employment for over four hundred people he added that his critics would do well to think again before smiling and ridiculing his achievements. It must in fact have been very difficult for any reader of the "Notice" to have ignored any of the better points of Emile Jumeau's character and business ability, as they are stressed repeatedly. He had, we are told, kept the honour, integrity and commercial reputation inherited from his father and added to this philanthropic works as well as his own energy and skill. Would not a national award to add to his numerous medals won at exhibitions be a suitable crown for his life?

Much to his delight, Emile did not have long to wait for, on December 29, 1885, he was granted his decree as Chevalier of the Legion of Honour and in 1886, giving his address as 8, Rue Pastourelle, was invested with the order.

It is difficult today fully to appreciate the effect that some of these very beautiful dolls of the late nineteenth century created. We immediately suppose that the public regarded the contents of toy shops much as it does now, simply accepting the quality as a fact of life. Contemporary accounts suggest that this was not so, and that the dolls were so immediately striking that the most sophisticated of foreign journalists and wealthy society ladies commented on the beauty of the French products in their reports and journals. The two main Paris toy shops in the 1880's were Au Paradis des Enfants

This is an example of an Emile Jumeau that has not passed through the hands of many destructive dealers and has retained its unusual plaited animal hair wig. This was originally arranged in a much more complex form but enough remains in place to show what a realistic effect the wig-maker was aiming at. The doll has a jointed body not, it should be noted, of the ball jointed type, though she has fixed wrists. Her original brown leather shoes are impressed "E.Jumeau" and she wears a contemporary ribbed lace type dress trimmed with ribbon and coarse lace. Her head is incised "Déposé E 8 J". The child type body is stamped "E.Jumeau Med.Or 1878 Paris". Height 18½ inches (47cms). Courtesy Abla Odell. Photograph Acanthus.

and Au Nain Bleu, and writing in the twentieth century of her childhood, the Baroness de Stoekl described the dolls of various sizes that were to be found on the shelves, all with beautiful wigs and exquisite little buttoned boots. Some were made of wax "dressed in long clothes of real lace, even the nappies could be embroidered with the coronet and initials of the purchaser. If modern children were taken there, they would remain with their mouths open in amazement". She also described how the Champs-Elysées was a kaleidoscope of colour because of the richness of the children's clothes and that of their wet nurses, who usually wore their provincial costumes and sat contentedly suckling their charges in the sunlight. The most fashionable children were frequently dressed in English style and the finest sailor suits came from a shop near the Opera called "Old England". The affection for sailor suits for both girls and boys is seen in Jumeau's products, especially those made in the 1890's when red as well as blue cotton was used for their manufacture. This most Parisian firm, however, avoided the tartan kilts that were also sold to the fashion conscious young at "Old England". A visit to this shop was made even more alluring as, just across the street was the famous sweet shop, Boissier. The Baroness, in wistful mood added;

> "When one remembers the luxury, the inexhaustable sums of money, gushing like a torrent from the rich quarters of Paris, it seems inconceivable that it lasted so long. Wealth did not seem to count in those days; millionaires enjoyed their wealth and those with small incomes or none at all, enjoyed the hospitality of the millionaires, so all was in order".

A new extravagance

A carelessness with money was something quite new in French bourgeois character and was probably fostered by an awareness that all could easily be lost, as it had been during the Siege of Paris. Professional people and well paid artisans joined the wealthiest sections of society in a willingness to buy well made and beautiful dolls for their children, and the toy shops offered an irresistible selection, ranging from the Jumeau walking dolls to his "Bébé Parlant", described in an 1885 advertisement as made in sizes seven to twelve. In the same year, a new series of dressed bébés was marketed "With a green arm-band of the so-called satin-réclame. Sizes 1 - 6". The term "Talking doll" was at this time something of an exaggeration, as they worked by the insertion of a simple voice box in the torso, activated by the pulling of a cord. In some instances this type of body was used in combination with eyes made according to the July 1886 patent, a method that was generally described in the "Notice" but formally registered in the following year.

The heads of the bébés made at this time were particularly lovely, with large luminous eyes that were accentuated by dark, almost lowering brows that would have seemed strange on an actual child but added character to the dolls' faces, that could too easily have become over-sweet in effect. Some of the most effective heads are those marked with Emile's initials, "E.J." together with the size. This type of doll, often on the eight ball jointed body, almost invariably has a good quality head, though there are exceptions, and "E.J's", like any other dolls, have to be judged on individual quality. One of the smallest sizes recorded measures only nine inches but some were large, the size twelve being 25½ inches. Although the finer "E.J's" are mounted on the early type bodies with fixed wrists, marked "Medaille d'Or. Paris", examples are also known that seem original but have the later body, again indicating that the firm continued to re-use heads over a long period. In his advertisements and publications, Jumeau

This country girl, wearing a striped serge frock and a cream silk pinafore carries a Steiff lamb. The fine bisque head is incised "E.J A" with the size "10". These "A mould" bébés by Emile Jumeau are particularly liked by collectors, as the quality is always so high. The early type eight ball jointed body is stamped with a blue "Jumeau Medaille d'Or" mark. The doll has fixed blue eyes and applied ears. The applied ears are a feature of many of the finest Jumeaux. Height 26 inches (66cms).
Courtesy Emma Berry Collection. Photograph Acanthus.

An unmarked Bébé Jumeau with fixed brown paperweight eyes, pierced ears and a closed mouth. The jointed body has a shaped adult-style torso stamped "Jumeau Medaille d'Or. Paris". The head carries the Jumeau tick marks. She wears the original claret woollen dress, embroidered and trimmed with black ribbon. There is a gauze underskirt and whitework underwear. The red leather shoes are marked "9 Bébé Jumeau Déposé". Height 18½ inches (47cms). Courtesy Phillips, London.

irritatingly makes no mention of the marking of the dolls' heads, though he returns time and again to the satin bands which were used to differentiate the types. As these bands were often either removed by the young owners or simply lost, they are of little help to the collector of today, though their presence, in combination with other evidence, does help in arriving at a more positive dating of an example.

Several other fine Jumeau models also date from the glorious 1880's, such as the Jumeau Triste, known somewhat inelegantly in America as the "Long Face". The head of this bébé is completely unmarked, except for a size number, but is particularly lovely, being more elongated than usual, giving a sad, wistful expression, that is most appealing. The Jumeau Triste is found on bodies marked "Medaille d'Or" and has fixed wrists, though there are some exceptions, probably created because of the replacement of damaged heads or even because a dealer has seen the advantage of changing around heads of the same size. As with all curious examples that do not fall into the regular pattern, some reservations must always be retained and it is virtually impossible to be categorical regarding all the products of a factory that worked over a long period. Such problems add interest to collecting and researching; the search for clues, the combining of information, and the fact that, in many cases we can never arrive at a positive conclusion, all add to the allure of the bébé and force the collector to judge each doll as an individual item.

There is less contradictory evidence regarding the so-called "Portrait Jumeau", as these are always of the earlier type, with the bodies carrying the Medaille d'Or stamp. These heads must not be confused with the later character-type portraits, as they have little resemblance to actual people, but again suggest an idealised child. The modelling and manufacture of the "portrait dolls" is more in the manner of the Parisiennes, less child-like than the later bébés, with comparatively smaller eyes, and made of fine quality pale bisque. The bodies of these classic examples are of the eight ball jointed construction with the usual fixed wrists. Again very few really small examples are

A very pale bisque unmarked Jumeau with an open mouth with moulded teeth. The bébé has stationary blue eyes and pierced ears. She wears the original wig and cream silk swansdown-trimmed hat. The body is of the later type, with jointed wrists and incorporated joints, that gave the undressed doll greater realism. The head is incised "3". She is seated in a slatted wooden mail cart of doll size. Height 23 inches (58cms). Courtesy Pat Birkett. Photograph Acanthus.

known, the smallest being the nine inch size "O". Some examples of this doll are found without any mark on the body, and in other cases it is so rubbed and faded that it is difficult to read. The heads of portrait Jumeaux are of an almost uniformly high quality and it was on work such as this that the reputation of the company was built.
The so-called "Almond Eyed" Jumeau dates from the same period as the portrait dolls, but can be differentiated by the somewhat elongated eyes, as can be seen in the photograph of the two dolls. Both these terms are of American origin, and have developed from the need of collectors to formulate a common language to differentiate the various types of heads. Jumeau never made any distinctions and simply concentrated on enthusing over the latest model.
The typical,somewhat later Bébé Jumeau, with its smoothly perfect face, large eyes and heavy brows, is generally thought to have been the creation of the sculptor Carrier-Belleuse. Much contradictory information has appeared concerning this involvement, that was mentioned on the labels of Jumeau boxes dating to the 1890's. Before this time, curiously, Jumeau made no mention of the sculptor's work, despite the fact that he was more than willing to boast of almost any facet of the bébés' production. Was this because Belleuse was not actively involved until the late 1880's or was the reason perhaps more subtle?
American researchers attribute the design of the bébé heads to Albert E. Carrier, who liked to be known as Carrier de Belleuse (1824-1887), a sculptor who was the darling of Paris during the Empire and who had worked as head of the Minton pottery school of design in Britain, and eventually became art director at Sèvres. Unfortunately, the only fact at our disposal is the box-lid reference, which could well indicate the involvement not of the father but the son, Carrier-Belleuse. It does seem unlikely that Jumeau would have omitted the "de" which the eminent artist affected, while the date also seems much more in line with the commercial work of the son, who lived from 1848 - 1913, and was much the same age as Emile himself. It is often observed that the Belleuse head can be easily recognised, but the prudent would note that the dolls have only what is reputed to be the Belleuse head; there is no documentation and the artist could well have been responsible for just one or even several different models. We can only wish that Jumeau had been much more explicit.

The Antwerp Exhibition

Emile was now at the height of his success, and continuing to win medals such as those at Sydney and Melbourne in 1879 and 1880 and a gold at New Orleans in 1884. One of the most interesting events at which the firm won an award was the Antwerp Universal Exhibition of 1885 and it was at this that the most coveted Diploma of Honour was gained, a mark that was used on the majority of dolls produced after 1885 and which was proclaimed in his advertisements.

> "Bébé Jumeau Diplôme d' Honneur. A unique award among all the French toys. All my naked or dressed dolls bear the mark of the firm. The dressed dolls of the satin réclame series wear a maroon arm band. The dressed dolls of the novelty satin series wear a red arm band (with the words Bébé Jumeau in gold letters).
> The bébés in a chemise also have a label (of satin) on the chemise, with the words "Bébé Jumeau" in gold letters and are also marked on the neck and on the body. The satin réclame

A slightly smiling bébé with its original fair mohair wig and contemporary pink muslin frock. Her shoes are marked "Bébé Jumeau Déposé". She has fixed blue eyes, a closed mouth and pierced ears. The head is stamped in red "Déposé Tête Jumeau" and carries black tick marks. It is also incised "8". The body, with jointed wrists is stamped in turquoise "Bébé Jumeau Diplôme d'Honneur". The doll is particularly interesting as it is accompanied by a trunk of clothes dating from the late 19th century to the 1940's when it was last played with. The oldest group of clothes includes a nightdress. Height 19 inches (48cms). Courtesy Polly Edge, Laugharne Doll Museum. Photograph Acanthus

series is only made in the first eight sizes. Refuse any doll that does not have this guarantee. Gold Medals 1879. Paris, London, Vienna, Philadelphia, Sidney, Melbourne".

Jumeau had exhibited his dolls at Antwerp as part of Class 37, Bimbeloterie. The report of this jury was of added interest as the reporter and secretary was Charles Pean, President of the National Union of Toy Merchants, founded in 1883 with Emile Jumeau as one of its first members. This union of sixteen manufacturers worked towards the limiting of German imports and staged displays as well as showing French products to advantage at Exhibitions such as this. A most dazzling array of French toys appeared at this event, though the organisers were disappointed that the German manufacturers chose to show nothing. The main section among toys was consequently bébés and lady dolls and their accessories, and the leading French makers in this area fought really hard for prizes.

The Union of the Makers of Toys of 36 Rue d' Hauteville, Paris, who displayed some thousand articles of Parisian manufacture, would have been awarded a Diplôme d' Honneur for its initiative in creating the first bureau of samples had not its president been a member of the International Awards Jury. In order to avoid the accusation of partiality, the jury placed the association Hors Concours. Three of the coveted diplomas of honour were awarded at Antwerp: among them was that of Jumeau of the Rue Pastourelle.

"This is the first time that such a high award has been given to a toy maker. It is also the first time that one finds oneself in the presence of such considerable effort made by an industrialist in such a relatively short time. If one thinks that it was in 1875 that Emile Jumeau took over the firm founded by his father in 1843. This firm had reached a turnover of 150,000 francs, which, for the period, was a good result which could not have been expected to be superseded, for this was the normal level. E. Jumeau succeeded in 1884 however, in taking this turnover up to nearly a million francs for both naked and dressed bébés. The firm of Jumeau has taken part in a great number of exhibitions and from the awards obtained by it, we can see its constant progress. In 1844, one year after its foundation, it exhibited for the first time in Paris where an honourable mention was awarded. We see it successively in London in 1851, Paris 1867, Vienna 1873, Philadelphia 1876, Paris 1878, Sidney 1879, Melbourne 1880, New Orleans 1884, Antwerp 1885.

In these various exhibitions they have obtained all the categories of awards, honourable mentions, bronze, silver and gold medals and, in this final place, the First Class Award. Such results could only be obtained by continual efforts, numerous improvements and by the installation of a model factory, which M.Jumeau has had built at Montreuil near Paris. It is in this way that he has succeeded in making products of the greatest perfection. The great merit of the Jumeau firm is that it is now able to make in its factories all the parts which constitute bébés; not only the body is made at the factory but also the head, wig etc. In order for such a method of working to succeed, considerable sales are necessary. Today, the making has been organised in such a way that the firm of Jumeau can supply to the makers of dolls, heads and wigs, which specialist makers used to supply and they were able to do this much better than the makers of heads who, not having kilns of their own for firing, are obliged to have recourse to third parties, which adds to the general cost. The firm of Jumeau is not only able to make its own heads but also to fire and decorate them in its own workshops.

Previously, the maker of bébés had to have recourse to specialists; he had only to bring together the various parts which go to make up a bébé and seek openings for the sale of his work. Today all that is changed; many do still work like this, but it is to the detriment of the work of their firm for, in order to do well and be able to sell a lot, one must unite the entire manufacturing process. When one thinks of the sum of effort that has to be poured out to make a bébé one is stupified."

The deep indents at the corners of the mouth give this bébé a slightly smiling expression. The head is stamped in red "Déposé Tête Jumeau Bte. S.G.D.G" with the size "8". There are also red and black tick marks. The wood and composition body with jointed wrists is stamped in blue "Jumeau Medaille d'Or Paris". She wears the original mohair wig and has fixed blue eyes and pierced ears. The original cream wool lace-trimmed frock is worn. It is decorated with pink silk ribbons. A spare pair of kid shoes are marked "Vandeput, Bruxelles 8". Height 19 inches (48cms). Courtesy Sotheby's, London. Photograph Acanthus

The writer then goes on to describe the Jumeau factory, an account that is reproduced as it does add some information to the "Notice".

"First of all the kaolin has to be macerated which gives the paste, then it has to be made to the required thickness by passing through a roller according to the size of the head, naturally the smaller the head the thinner the paste.

The paste after being rolled is then cut into squares which are then put into two part plaster moulds, taking great care to support the mould so that the head is perfectly shaped. These moulds can only be used fifty times, after which they become useless. The heads are taken out of the moulds and then the eye sockets are cut out using a special tool and taking care to thin the inner part of the socket so that the eye fits better. Ears are made in this mould in the smaller and medium size heads and in larger sizes moulded separately and subsequently fixed. The moulded head is placed in a sort of sieve called a saggar, the head is covered with a sheet of raw kaolin so that the head does not directly touch the ground. A saggar holds about three dozen heads, for they are put one inside the other. The essential is that they do not touch. The saggars, when filled, are put into an immense kiln where they are fired for about twenty seven hours at a very high temperature. They are then taken out after having been allowed to cool, and when removed, the heads are transformed into white porcelain. Each head is rubbed smooth with glass paper and polished and then it goes for decoration which, at the Jumeau factory, is done either by men or women.

Decoration consists of two stages. The first, where it is given the ground colour and the second where the colouring is perfected and finished. Mineral colours are used for this purpose because the decoration holds better and is finer. This is the method by which it is done: two coats of pale pink paint are first put over the face, neck and forehead in order to give the head natural skin tones; after this, it is allowed to dry and then paints are again applied to the head, colouring the cheeks, then the lips, the eyebrows, eyelashes and nostrils. This work requires great attention, because so little as a finger sliding over a part of the face not yet dry and the head is lost and everything has to be started again.

When the decoration is finished, the heads are fired in a lower temperature kiln for seven or eight hours to be finished. Then the eyes are put in place. We will pass over the making of the eyes, which is in itself complex but only forms part of the face, as well as the making of the wig, so that we come to the unbreakable body.

The body is made with paper called *papier goudron*. Sheets of this paper are sandwiched with layers of glue size and laid on top of each other until the layers have the thickness required for the various parts of the body. Thus the back and chest need more solidity than the legs. This paper is put into heated moulds and is transformed, by the time it is taken out, into the trunk of the body.

One has to ensure that the edges of the body are well designed so that they will fit together and be solid, for the body is moulded in two parts which form the whole.

When the different pieces forming the body have been moulded, they are put to dry on racks and according to the season of the year, this requires one or two days. These then go on to be glued and for the insertion of the little pieces of metal which are used for stringing and which aid the articulation of the arms and legs.

This being done, the bodies are sent to the painting shop where they receive a coat of white, are then allowed to dry and the bodies rubbed down. Five coats of paint are applied and one of varnish and it is allowed to dry before being passed on for assembly.

The elastic of the leg passes into the hip and then goes to join to the body where it is hooked into the chest. When the body is finished a hook is placed inside the neck which is then tightened and holds the head in place. A piece of cork is attached to the head for the fixing of the wig. The bébé is finished. It remains to dress it. The bébé is sold both undressed, that is to

A most charming figure in the image of a child by Emile Jumeau. She has fixed blue eyes, applied ears and very well painted lips with a line of pale colour left between to give the illusion of a separation. She wears the original brown animal hair wig. The body is of the ball jointed type with fixed wrists and is stamped in blue "Jumeau Medaille d'Or Paris". The head is incised "8 E.J" for Emile Jumeau. The shoes carry the mark "Au Bébé Bon Marché Bocher, 63 Rue de Sevres". She wears the original lace over pink silk costume which was further decorated with matching pink satin ribbon. The sleeves are pin-tucked and fastened with buttons. The amount of work on this costume was considerable and can be compared with the much more simply achieved effect of that on the Jumeau Triste from the same collection. Height 22 inches (56cms).
Courtesy Abla Odell. Photograph Acanthus.

say not completely naked but dressed in a chemise, and dressed, in a frock of silk, wool or satin. As can be seen, the making of a bébé requires numerous processes. M. Jumeau does all these in his factory and has succeeded in making his mark throughout the whole world. Today, beautiful German bébés are sold in Leipzig under the name of Bébé Jumeau, because he has succeeded in making such a fine product under this name. Is not this imitation the consecration of the results obtained?
Thus the Jury has awarded to M. Emile Jumeau a Diplôme d'Honneur. We have devoted a considerable amount of time to the Diplôme d'Honneur to explain the motives which made the Jury award these high distinctions".

The Diplôme d'Honneur Period

It is very obvious that at Antwerp the competition among French toymakers was much greater than usual, so that the Jumeau award was obviously richly deserved. The Jury complained that the German exhibitors had not replied to the questionnaires they had been sent, so little could be published regarding their products. Jumeau, being so publicity conscious, was obviously very willing to provide both organisers of exhibitions and journalists with an abundance of information regarding the making of the bébés, and it seems from the lack of variety between journalists' reports that they were supplied with much of the content by Jumeau himself and simply added a few touches of their own. Gaston Tissandier, writing in "La Nature" in 1888, was apparently taken on a tour of the factory by Emile Jumeau and described the establishment as the largest doll factory in the world, producing some figures that stood as tall as a four year old child. He also described how the women who cut the eye sockets made use of a mould to make sure that they would be in exactly the right position. Jumeau mentions this use of a template in his patent application and it seems that before the mid 1880's the process had been less tightly controlled and left more to the basic skill of the workmen. Tissandier supplies the information that the mechanisms for these new eyes, patented in 1886, were assembled in the workshop where the male workers usually fixed the eyes in position. He also adds that the first of the five coats of paint on the body was zinc white. In a most interesting illustration, he showed the machinery that was used to fix the very characteristic neck spring. A close examination of the print included in the article reveals that the doll was held vertically in a wooden frame when loosely strung and the head stringing loop was put under tension by being attached by cord to a chain, while the spring was put under opposite tension being held down by a goose-neck lever. The operator appears to be picking up the short metal rod that was used to hold the metal stringing loop against the spring. The use of this spring was presumably intended as a means of taking up some of the tension as the elastic weakened. Its complexity, in comparison with the much simpler German methods, is curious and we have to assume that it was Jumeau's affection for engineering processes that encouraged him to use an expensive, capital-intensive method of stringing when he could have cut costs by using the simpler method utilized in many Parisiennes. If only suitable equipment still existed, these springs could still be used for the re-stringing of dolls, as the condition has usually remained good, though few collectors would now like to risk cracking the head just in order to string their acquisitions in the classic Jumeau manner.

A striking pair of closed-mouth Jumeaux, the bébé in dark blue having blue fixed eyes and a fully jointed body with articulated wrists. The head is stamped in red "Déposé Tête Jumeau 8" and also incised with the size number "8". There are red and black tick marks. She has the usual pierced ears. Height 20 inches (50cms). The doll with the abundant blonde wig has brown fixed eyes and a closed mouth. She has fixed wrists. The head is stamped in red "Déposé Tête Jumeau S.G.D.G. XIII 8 T". Height 19 inches (48cms).
Courtesy Terice Tipper. Photograph Acanthus.

Detail of a Bébé Jumeau label found on an 1890 box. (Translation in text).

The dolls of the Diplôme d'Honneur period show Jumeau's skill and ambition at its height. The quality of the heads was good and further improvements in the construction of the body had resulted in a more realistic figure, that had evolved from the incorporation of the original separate ball joints into the basic structure. At first the original eight was reduced to four, at elbow and knee, and after the Diplôme d'Honneur, this was further simplified so that all the separate turned wooden balls were dispensed with completely, creating a doll that looked more like an actual child when undressed. He was also very active at this time in the promotion of his company through articles, booklets and many advertising exercises, such as the various coloured ribbons which his dolls wore. Strangely, for the size of the firm, few patents were registered, and one, dating apparently to 1862 and mentioned in several recent accounts of the firm, simply does not exist, though a completely different company with the name of Jumeau did register a patent in the year mentioned, for an improvement in paper. Presumably this name was picked up by some early researcher and perpetuated in accounts of the company, though there is no factual basis and the firm only seems to have become involved in the registration of designs after Emile Jumeau took control.
His patent for moving eyes, Number 177127, dated July 1st 1886, is complex even in the extract form used by the patent office, but does explain, in Emile Jumeau's own words, exactly how such eyes should be activated. The design was intended as a means of moving the eyes. It was claimed to be both strong and cheap to produce. The

This large closed-mouth Jumeau sits comfortably in a Victorian child's high chair with its safety rail forming a good book-rest. She has fixed blue eyes, pierced ears and a black real hair wig. The body is of the later jointed type with articulated wrists. The head carries a red stamp "Déposé Tête Jumeau 12" and has additional black tick marks. She wears a Victorian whitework lawn lace-edged frock with a ruffled organdie bonnet. Height 27 inches (69cms).
Courtesy Lilian Middleton's Antique Doll Shop. Photograph Acanthus.

workpeople no longer cut out the eye sockets in the traditional way but were given a template so that the shape was always exactly the same, eliminating the possibility of error so that the eye would always turn smoothly. (See appendix 1.)
This patent specification is a single instance of Jumeau's great attention to detail and also indicates how the complete process of manufacturing a doll was becoming much more tightly controlled, leaving less to the skill of the individual workman. After 1885, there is also a dramatic increase in the amount of documentation regarding the firm's products, partly because the dolls sometimes carried informative advertising material on their boxes and because of the booklets, the game, and cards connected with the various types of doll. In 1886 Emile registered his mark Number 24407 for "Bébé Jumeau" described as "A mark intended to be put on the dolls that I make as well as the wrappings, boxes etc. for the said dolls. The characters of this mark will be of a size and colour appropriate to the dolls. They will be gold on satin and placed in all suitable places". This trademark was also registered a few years later, in 1888, in the United States. It seems likely that the satin labels were only used from around this date and I have certainly never examined any dolls in original costumes that can be dated earlier than this which carry these satin marks. In advertisements of the 1880's and 1890's these Jumeau marked bands are much in evidence, even in the simplest line drawings, and were fixed to the waist, hip or arm of the bébé. On the back of the illustrated lottery ticket, issued in 1886, Emile Jumeau again took the opportunity to make the French child completely aware of the superiority of his product.
The front of the card showed two dolls pointing to a photograph of the Bébé Jumeau factory, the larger version revealing the separately articulated wrist, those of the smaller still being of the fixed type. The big doll, with her elaborately curled hair, is dressed in the typical outfit of the 1880's, with a dropped waistline and feather decorated hat. Of particular interest is the fact that in this case the satin band was worn around the upper arm. The small bébé, wearing a lace-trimmed chemise, carried the label near the hem. On the back of the card we read:

> "Bébé Jumeau Diplôme d'Honneur. A unique award achieved only by this doll up to the present day among all French toys.
> Le Bébé Jumeau.
> With a pretty head in bisque, long and silky hair, enamel eyes, unbreakable body which enables it to make all natural movements and to assume all positions, a hat with feathers in the latest fashion, the dress of wool or silk designed by the leading Parisian costumiers, shoes of gilded kid, socks of pink or blue for the day, gloves of suede, a necklace, earrings. Always smiling, adorably pretty, such is the Bébé Jumeau.
> Nothing has been neglected to make this doll the most beautiful, the most complete, the most enjoyable of toys; it is the present par excellence, for our young girls, small or large, with which they will never be bored; it can be dressed, undressed, hold out its arms, kneel or sit; it is a model of obedience for its little mother.
> The Bébé Jumeau does not break; it is eternal; if its beautiful dress becomes a little worn, one can make another for it and it will last again for a long time.
> It stands firmly on its own feet and holds out its arms to whoever approaches; if it could but speak what then! If you wish, it can be made to say Papa and Mama in a high and intelligible voice. Therefore be sure of its identity and do not buy the bad imitations. The Bébé Jumeau has a passport; it is marked; it has its name on the body and when it is dressed, on the left arm. It is Someone and the law wishes you to respect it.
> Bébé Jumeau has friends in all classes and levels of society, for the Bébé Jumeau is well brought up. It has travelled to London, Vienna, New York, Milan and Madrid, all the capitals of the civilised world have been visited by it, and in its mother country it has known how to

The beauty of the open mouth Jumeau is often under-appreciated by collectors. In this photograph, the quality of the bisque and the assured painting of the lashes and heavy eyebrows can be studied in detail. This is an umarked Jumeau, but few buyers would have encountered any difficulty in identifying work by this leading French maker. The blue eyes are fixed and the moulded ears pierced. The doll has the later type Jumeau body with jointed wrists. The head is incised with the size "10". She wears a replacement French wig. Height 23 inches (58cms).
Courtesy Lilian Middleton's Antique Doll Shop. Photograph Acanthus.

attract general affection for Paris idolizes it. Lille, Roubaix, Nantes, Nancy, Strasbourg, Metz and Mulhouse have adopted it and in each year of this period have given it the most marvellous reception in the whole world. All our little French girls should have a Bébé Jumeau as a Christmas present this year. The Bébé Jumeau is the true French doll, the little graceful Parisienne who is as pretty as possible, who brings joy and satisfaction of the most beautiful dreams of Christmas to the children of France.

Happy the child who can possess it.

No.24686. This number gives a right of entry into a lottery for a dressed doll which will be drawn on December 20,1886. The winning number will be published in the Petit Journal."

The fame of the Jumeau name was carried to many people in the catalogues of shops such as "Les Grands Magasins du Printemps" which, in 1887, was offering fully articulated Bébés Jumeau with bisque heads. In addition to this basic advertisement, "Les Grands Magasins du Printemps has the honour of giving notice that from December 6 the Bébé Jumeau will be sold with great reductions on the previous price list." Such sales ploys were constantly used by Jumeau, in this instance probably because they were eager to dispose of dolls made surplus to Christmas requirements, which were sent off in October of each year. The Printemps catalogue was not only sold in France but was also widely advertised in Britain, giving mothers an opportunity to buy an authentic French doll as a Christmas present. The bébé illustrated in 1887 was shown without shoes and wearing the typical, low-waisted frock of calf length with the "Bébé Jumeau" label on the hip. It was available dressed only in a chemise, in a silk chemise, in various silk dresses or, the most expensive, in various outfits in rich silk. In the ordinary grades of costume the dolls were not sold dressed above 48 cms, while the most expensive costume, in rich silk, was not made for the seventy centimetre size, this being the largest offered.

Many of the claims made in Jumeau advertisements serve only to confuse, as the collector wallows in contradictory evidence regarding marks on bands, heads and bodies, and is really best advised to rely only on the evidence of the doll itself. Even here, we are not always on safe ground as it has to be remembered that the firm was willing to re-dress dolls, that families often brought the clothes up-to-date, and that, even more confusingly, toy shops frequently supplied new heads for damaged toys. In a late 19th century advertisement from the London toy shop, Hamleys, new heads were offered in a separate section of the Jumeau page and ranged in price from a size one which cost 2/- to size fourteen which cost 14/6d. This price included the fixing back of the head to the doll but the original wigs were re-used. Hamleys described themselves as the special London agents for Jumeau dolls and stated that the First Quality Bébé Jumeau was the highest quality doll made. The number seven size, measuring 17 inches, sold at 14/6d, while the thirty inch, size fourteen was offered at the high price of 54/6d. All the sizes were not available in the talking Jumeau dolls, which said papa and mama, but only sizes seven, nine and eleven, the latter costing just under two pounds. The second quality Bébé Jumeau, made in exactly the same sizes as the premier description, were sold in smaller sizes, the Number 3, measuring twelve inches costing under six shillings. These second quality dolls, with flowing hair and wearing shoes and socks, were available only up to size eleven. The second quality bébés were also sold with the talking mechanism and sleeping eyes, also in sizes seven, nine and eleven.

An 1887 advertisement stated that the firm had sold more than 130,000 Bébés Jumeaux and in this same year, at the height of Emile's success, another patent was

Purchased in Mons, Belgium, and known as "Amandine" this large open mouth Jumeau wears a contemporary child's whitework and broderie anglaise frock and cape. She has a fully jointed body stamped in turquoise "Bébé Jumeau Breveté S.G.D.G Déposé". She has fixed eyes and moulded teeth. The head is stamped in red "Déposé Tête Jumeau 12" with tick marks and a decorator's mark, "19". She has the original animal hair wig. The body contains a pull-string voice mechanism. The holes for the insertion of the string being strengthened with brass eyelets. As with many German dolls, the body was simply cut in two for the insertion of the mechanism and the join secured with a strip of plaster. Height 27 inches (69cms). Courtesy Polly Edge, Laugharne Doll Museum. Photograph Acanthus

1890 **CHAUSSURE** DU **BÉBÉ JUMEAU** 1890

Le **BÉBÉ JUMEAU** est vendu chaussé : Sa **CHAUSSURE** est de première qualité ; elle est faite d'un SEUL MORCEAU pris dans le cœur de peaux de 1er choix : (et non en deux : pris dans des déchets ou rognures de peaux qui s'achètent en solde.) La SEMELLE est en cuir naturel (et non en carton recouvert de basane.) C'est, en un mot, une chaussure supérieure venant compléter le BÉBÉ JUMEAU déjà supérieur lui-même à toute autre fabrication.

MÉDAILLE D'OR -- DIPLÔME D'HONNEUR : HORS CONCOURS EN 1889

il a donc obtenu toutes les récompenses possibles

A L'EXPOSITION de 1889
la vitrine des BÉBÉS JUMEAU
a été visitée par plus de SIX MILLIONS de personnes

(Succès sans précédent dans les Annales du jouet.)

N. B Le **BÉBÉ JUMEAU** devait paraître tout chaussé le 1er Janvier 1889, par des circonstances indépendantes de la Direction, il n'a pu l'être qu'en 1890.

The 1890 Jumeau shoe declaration found on a doll box of the period. (A translation is found in text).

taken out for sleeping eyes that in this instance, included eyelashes. This patent, Number 182307, registered on March 21, 1887 was again described as intended for improvements in the eyes of dolls "or others" and gave them a sidways movement. It also moved the lashes by the motion of the eyes. (See appendix 2.)

This 1887 patent, that was attempting even greater realism, seems, despite its mention in advertisements, hardly to have been used and I cannot recollect having seen a single example, though a number of those with the earlier mechanism, of which this was an improvement, are found. The mechanism used for these bébés was not simple and could not have been anything like as inexpensive as Jumeau claimed in his patent specification and it seems probable that the production remained fairly minimal. The metal fittings in the heads also meant that a doll, if dropped, had less chance of survival, and even among the known examples, many are found with cracks.

Another improvement of the late 1880's was the introduction of the open-mouthed Jumeaux with teeth, the first reference to these dating to 1888. The open-mouthed doll was considered a great advance by all the dollmakers of the period and naturally cost more to produce than the traditional head with its closed lips. Simply because this doll

A pale bébé Jumeau wearing a very light pink lace-trimmed frock. She has fixed blue eyes and a softly painted closed mouth. As is usual, the ears are pierced. She has the late 19th century body with jointed wrists. The head is incised with the size "8" and has black tick marks. Despite the doll's high quality, the body is also unmarked. These unmarked dolls lose little in value because of their lack of a complete mark, as knowledgeable collectors value a Jumeau on its quality and general effect rather than on the presence of a stamp. It is also worth noting that some of the most attractive dolls are unmarked while some very poor specimens carry full details of the maker. Height 19½ inches (49cms). Courtesy Dawn Herrington. Photograph Acanthus

continued to be made in vast numbers well into the twentieth century they are less popular with collectors, who prefer the rarer dolls. Purely on aesthetic grounds, the closed-mouthed dolls are more satisfactory, as the open mouth never looks really natural and one is always aware of the gaping emptiness behind. Jumeau's open-mouthed dolls of the 1890's were much more tasteful than some of the later versions, but are considerably less attractive than the earlier products. Originally the open-mouthed dolls sold at about four francs more than the traditional type.
Some concept of the development of the factory is given in the lists of employees which Jumeau was fond of publicising. One states that in 1885 there were five hundred, in 1886 eight hundred and that the number had risen to a thousand by 1890. The number of dolls also increased dramatically from 10,000 in 1881 to 3,000,000 in 1897, the latter figure including white, mulatto and black heads, the coloured dolls having been introduced in 1892.

The costuming of Bébés

The introduction of the bébé, essentially a child's doll that was intended to be played with and undressed, seems to have inspired Madame Jumeau and her small army of needlewomen to even greater heights, and it is particularly noticeable that the underwear of the dolls was much improved, as many Parisiennes had worn very economical petticoats and drawers. Aware that the child would wish to change the bébé's clothes frequently, the dressmakers added generous layers of lace, pleated frilling and embroidery to the underwear. After 1880 a type of permanently pleated muslin, edged with lace, became almost a signature of the company and was particularly useful in holding out the low waisted calf length frocks that were so fashionable. Some of these petticoats have unusual V-necklines, which were complemented by the same shape of the chemise beneath and were obviously made very specifically for each size, as the fit is always exact. The underwear is almost invariably machine sewn, but the complexity of many of the pleated and ruched frocks meant that they were in fact much easier to assemble by hand. The size of the stitches on some of the dresses, which present an eye-catching effect, is very large, with some being as long as an inch, making the wrong side of an outfit extremely untidy. The skill of the costumier, however, was such that the finished effect looked both complex and beautifully executed and made the French doll stand quite apart from all its rivals. It was this concentration upon a purely superficial effect that caused the English and German writers to be highly critical of the bébés, and to compare their slightly vulgar outfits with the plain neat work of their own, somewhat staid, dressmakers who offered children dolls that were costumed in the simple dresses they wore themselves.
The French child was herself something of a curiosity to children of other countries, who looked at her rich clothes in amazement, being very aware that in their own rough and tumble, such fine satins and silks would be ruined instantly. The French bébés were therefore in themselves objects of great attraction, and stood in the toyshops, completely apart from the much plainer German products, making the youngest child aware that here was something really special. It was on this fundamental difference of approach that Jumeau based much of his advertising, that aimed at encouraging the child to want a doll made by his firm in preference to any others and it was upon such a basis that his Eiffel Tower game was printed and used as part of the packaging of dolls

A game of croquet. The bébé wearing the fur tippet has fixed blue eyes, pierced ears and a closed mouth. The head carries red and black Jumeau tick marks and the original Jumeau shoes are worn. She wears the original white piqué coat trimmed with lace and a silk hat. Height 20 inches (50cms). Author's collection. The bébé wearing the printed cotton dress under her broderie anglaise coat has fixed brown eyes and a closed mouth. She wears the original real hair wig. The head is stamped "Tête Jumeau" in red and the body is also marked. Height 21 inches (52cms).
Courtesy David Barrington, Yesterday Child. Photograph Acanthus

that were dispatched to America in 1889, the Eiffel Tower having been constructed to illustrate the potential of steel for the International Exhibition held in Paris that year. The rules of the game, in which a Jumeau holding a tricolour and a German doll carrying her national flag compete, are decidedly partisan, imparting the initial information that "Jumeau's dolls bring good luck, we therefore hope young players may not be so unfortunate as to be assigned places where there are German dolls."
Jumeau dedicated his game to young Americans, who were informed that the number of players was not limited and that it was played by throwing dice.

> "But players must not stop at the Jumeau dolls beating noisily on the drum. When the number points to a Jumeau doll, go as much further on, repeating if necessary until you stop elsewhere than on one of them.
> The first to reach 63 wins the game. If on nearing 63 the next throw takes you above that number, you are to turn back as many points as you are above it.
> Should you turn up 0 at the first throw, by a six and a three, your place is number 26 (where Jumeau's doll is hand in hand with young America). If you attain the same result with a four and a five you go the 63 by multiples of 9, by doubling successively you reach 63 where you see Jumeau's doll holding the French flag and triumphing over its rivals, and win the game.
> Whoever turns up six at the first throw, where the German doll is seen upon a bridge, must pay 1 and place himself at No.11 where such a doll falls into the water and gets soiled.
> Whoever goes to No.19, Jumeau's doll factory, must pay 2 and stay there until his partners have each played twice.
> Whoever falls into the well (i.e. No.31, where the German doll is placed,) must pay 3 and wait until another player comes and takes him out; he will then take the place of such a player. The player whose lot it is to go to No.42, where Jumeau's doll has lost its way, must pay 2 and go back to No.30 where the said doll has found the road to France.
> Whoever is assigned to No.52, where the German doll is in prison, must pay three and stay there until another's bad luck brings him there: the former then takes the place just left by the latter. The player who gets to No.58, where the German doll is seen broken, must pay 3 and go back to No.1. Any player who is met by another must pay 1 and go back to the place previously occupied by the latter.
> The fine large factory at Montreuil, where Jumeau's dolls are made, was only a small building at first but has gradually enlarged to meet the great and ever increasing demand for these popular dolls.
> It now stands on 6,000 square yards of land, and finds employment for upwards of 500 hands.
> All Jumeau's dolls are stamped with the maker's name and may, moreover, be distinguished by that inimitable taste and finish peculiar to French goods.
> The rules for playing this game are the same as for the Game of Goose, the only difference being that the figures are changed. We hope that you will get this game boxed up and preserved as a keepsake of the Great International Exhibition held in Paris in 1889 and as a token of the Franco-American Union".

The game must have been a great success, as copies are now extremely rare but are of considerable interest, as the somewhat spiteful incidents and phrases bring to life the very real enmity between the two countries and the almost child-like manner in which Emile Jumeau was prepared to take every possible advantage over his rivals. At the 1889 Exposition mentioned in the game, Jumeau was a member of the International Jury and therefore, although he exhibited dolls, Hors Concours. This event was a fiasco in relation to the toy section as Jullien, also a doll maker, was the first reporter for the section but died without producing a report, adding to the problems by leaving no notes. Charles Péan was given the task of producing some sort of record and obviously felt great chagrin because of the lack of information, especially as his own report at Antwerp had been unusually good. He informs us that 115 French makers displayed

Seated in a small doll's pram, this delicately coloured, closed-mouth Jumeau reflects the idealised child that the company strove to reproduce in bisque. She has fixed eyes and pierced ears. The body is of the "improved" type with jointed wrists. The torso is stamped in blue "Jumeau Medaille d'Or Paris". The head carries a red "Déposé Tête Jumeau Bte SGDG" stamp, together with the artist's ticks and a red stamped size "6". Height 15½ inches (39cms).
Courtesy Pat Birkett. Photograph Acanthus.

The completely original, shop condition, open-mouthed Jumeau is desirable to any collector. This example carries its waist ribbon and its original shop price label. This flowered cotton was used for a large number of dolls while the pleated effect was also very popular and used on Jumeau underwear.
Courtesy Musée des Arts Décoratifs. Paris.

toys but as the Germans sent no exhibits there was in fact little international competition. Many of Péan's comments are based on those he made at Antwerp but he does add that "We can, without exaggeration, boast about the manufacture of Parisian bébés. Our French factories are known over the world, and we can say that each of our makers has borne his personal part in the success of the bébé. They are not only well made but cheaply"; perhaps a comment that was directed at Emile, who seemed to claim the whole development of the bébé himself.
The inventions displayed at the 1889 Exposition included a phonograph mechanism shown by Edison, an idea that appealed to Jumeau, as he saw its potential in a miniature form, as a speaking voice for a doll that could surpass all competition and be

Two chubby and unusually childlike bébés Jumeaux. Both have fixed wrists and sturdy ball jointed bodies. The smaller seated bébé wears the original red silk tunic over a red and white dress. The shoes are marked "Bébé Jumeau Déposé". Her bisque head is incised with the size "3" and the body is stamped "Jumeau Medaille d'Or". Height 13 inches (33cms). The standing child, also in original costume, is marked only with an incised "l" on the head. She has fixed blue eyes and a closed mouth. The original wig is still in fine condition. Height 16 inches (41cms).
Courtesy Abla Odell. Photograph Acanthus

an object of wonder and curiosity. Even the finest dolls he was able to offer in 1889 would be as nothing beside the doll-maker's ultimate dream, the vision of a speaking figure. The company was obviously in a good position to make such an advance, as Jumeau stated that they were employing nearly one thousand people and in 1889 made 300,000 dolls. "It is a success without precedent in this branch of French manufacture". He provided this information in one of his advertisements, which showed a print of a child playing with a doll drawn with such unnecessary realism, apart from its stiff legs, that it bears little resemblance to any of the firm's products. The advertising material was completely in the Jumeau style, boastful and self-congratulatory, and presumably intended as much for parents and retailers as for the child.

> "Bébé Jumeau Diplôme d'Honneur. Unique award amongst all French toys. An essentially Parisian make. The doll preferred by people of taste. Monsieur Jumeau asks us to inform the public that the immense success of the doll has been achieved despite counterfeits being made in different areas and especially Germany.
>
> To avoid being tricked by certain sellers who are often only looking for a *big profit* you have only to insist on the Jumeau Mark, which is found on the back of the doll, on the chemise and on the box. When dressed, it wears an arm band on the left arm with the same inscription. Having said this, if you wish to judge the importance of the firm you have only to compare the following information: in 1879 the House of Jumeau with 30 people made 10,000 dolls; in 1889 the House of Jumeau will employ nearly 1,000 people and will make 300,000 dolls. It is a success without precedent in this branch of French manufacture. Above all, insist on the name".

The dolls of this period were certainly of fine quality and the general standard was much more uniform than it was to become in the twentieth century, when the firm of Jumeau had lost most of its individual control. The heads produced in the 1890's offer the collector a wide selection, as, judging from the evidence of original costume, it seems that several of the older types continued to be made alongside the more conventional bébés. It was also in 1892 that the first mention of coloured bébés made by this firm is found, though by 1895 there are many references, suggesting that the new type had proved very popular and was, of course, especially suited to the children in the French colonies and America. The number manufactured for the home and British market was much lower than that of the U.S.A., and consequently examples are found more easily in America. Some of the coloured or mulatto bébés were costumed in outfits similar to those worn by the traditional bébés, but others wore variants on national costumes. Curiously, these were rarely made as well by the dressmakers, and the finish was often weak in comparison. Simply because many of these outfits seemed somewhat tawdry to collectors of the last decade, the original clothes were sometimes stripped off and replaced by what were felt to be prettier substitutes, consequently depriving the figure of much of its historical significance. Fortunately collectors are becoming more aware of their personal responsibility in the preservation of antiques and there is no longer such a willingness to dispense with the characteristic features of dolls.

Madame Jumeau produced some highly ingenious ensembles for the attractive dolls her husband made in the 1890's, some of the more expensive silk and satin dresses being extremely complex. A few fit so tightly that it is almost impossible to remove them without damage, even though originally the hooks and bars were completely functional. Silk and satin shoes were also carefully selected exactly to complement the frock, as well as lace socks in suitable shades. In 1891 Emile registered the bee mark (No.36987), reminiscent of the glory of France in the Napoleonic era. This trade mark was cold stamped onto the shoes and was accompanied by the size and "Paris Déposé". It is a very characteristic mark but cannot always be relied upon as an indication of date,

A softly attractive closed mouth bébé with fixed brown eyes, pierced ears and delicate colouring. She has a jointed body with fixed wrists, stamped in blue "Jumeau Medaille d'Or, Paris". The head is stamped "Déposé Tête Jumeau 8" in red and is also incised with an "8" for the size. She wears an original faded white dress printed with a pink flower pattern and probably original. Height 19 inches (48cms). Courtesy Sotheby's London. Photograph Acanthus.

Though the lace which Jumeau used has lasted well, the fine silk and chiffon seen in many of the most complex costumes has frequently disintegrated. This charming example with jointed wrists and a later shaped body has a closed mouth and wears pale turquoise blue. The doll illustrates the effectiveness of a Jumeau costume, even in a size as small as this. Height 13 inches (33cms).
Courtesy Sotheby's, London.

as shoes were often replaced at many of the shops that specialised in Jumeau products and several dolls that are certainly earlier are found with the bee mark shoes that fit so perfectly that they must have been supplied by the firm. The earlier shoe mark had included the size, and "Bébé Jumeau Med. d'Or. 1878. Paris Déposé". One advertisement included details concerning the shoes and declared;

> "1890 - Bébé Jumeau shoes - 1890.
> The Bébé Jumeau is sold shod. Her shoes are of the first quality. They are made of a single piece taken from the centre of first grade skin; (not of the pieces taken from the throw-aways or off-cuts sold by workshops). The sole is of natural leather (not card covered with cloth), and is in a word a superior shoe completing the Bébé Jumeau, which is itself already superior to all other makes. Medaille d'Or, Diplôme d'Honneur; Hors Concours 1889. It has thus obtained all the possible awards. At the Exhibition of 1889 the showcases of the Bébé Jumeau were visited by more than six million people (a success without precedent in the annals of toys). n.b. The Bébé Jumeau ought to have appeared shod from January 1, 1889 but by circumstances independent of the management this could not happen until 1890".

Many of the Jumeau shoes are quite beautiful, with their characteristically pointed toes and realistic leather soles. Both the silk and leather shoes were lined most carefully with calico and all the edges correctly finished so that they were, in effect,

A pair of later Jumeaux, the larger wearing a contemporary child's whitework cape. She has an open mouth, fixed blue eyes and jointed wrists. The head is stamped in red "Tête Jumeau" and incised "12". The body carries the oval cream label with blue printing "Bébé Jumeau Diplôme d'Honneur". She stands 28 inches high (71cms) and carries a small doll incised "1907" with fixed dark blue eyes and jointed wrists. These were produced after Jumeau's amalgamation with other dollmakers. This is a size "6". The quality of the head suggests that it was one of the "1907's" produced in Germany for use on French bodies. It stands 15½ inches (39cms).
Courtesy Dorothy Brooke. Photograph Acanthus.

absolutely correct miniaturisations. Some were decorated with rosettes made of ribbon or leather, others have white metal buckles. A few dolls intended to represent characters in fancy dress or in national costumes wear slip-on shoes with rosettes but the vast majority represent the footwear of children, with straps or neat laces. The presence of the original footwear enhances the value of any Jumeau quite considerably, as does an original arm-band or label and one of the most upsetting features of the auction rooms is the frequency of thefts of these small items that appear insignificant but add such a touch of authenticity. The colour of the shoe decoration or ribbon is sometimes reflected in the edging of a coat or by the collar of a dress, all giving the bébé that distinctive Parisian touch.

The Jumeau ensembles sometimes take the form of a ribbon and lace-decorated frock and at others, represent a coat worn over the dress, though these more complex designs can sometimes be removed in a single operation. Elaborate buttons were a popular device for ornamentation but rarely serve any functional purpose, as hooks and bars were almost invariably used for fastening, as they could be concealed and were so much neater. Occasionally edging panels of flowered brocade combined with plain satin and edged with braid are seen on some of the more expensive outfits, detail that was extremely labour intensive and required skilled, artistic dressmakers rather than the very mediocre sewing women who made clothes for the majority of British and German dolls. Contemporary catalogues show that the dressed Jumeaux were always more expensive than the dolls costumed by their rival, Bru, and the House of Jumeau was obviously very aware that sales depended quite heavily on this important aspect. It is a feature which is too often lost today in the valuation of dolls, as a re-dressed Jumeau or even a doll without clothes will often sell at a figure comparable with a completely original example, indicating how the emphasis is now, so often, completely on the quality of the head.

The Jumeau dressmakers made knee and calf length skirts that were pleated, tucked or shirred, the hemlines rising in the 1890's to just below the knee. Cheaper dresses were made of fine cotton and a few were costumed in crocheted outfits with contrasting bands of colour, a style that must have made the bébés very pleasant to handle because of their softness, but much less attractive to those who seek the hall-mark of the firm in the costume. The hats were also given considerable attention, and some complex designs built up over card or buckram shapes. These smart bonnets and hats were usually lined and were often further decorated with feathers or cockades, one having small birds made of green feathers that complemented the colours of the outfit. On a box dating to 1890, Jumeau again stated that all these finely dressed bébés were marked,

> "Bébé Jumeau. Diplôme d'Honneur. A unique award among French toys. All my dressed and naked dolls bear the mark of the firm. Dressed dolls of the series satin réclame bear a maroon arm band. Dressed dolls of the series Satin Nouveauté wear a red arm band (with these words, "Bébé Jumeau" in gold letters). Dolls in a chemise also have a satin label placed on the chemise with the words "Bébé Jumeau" in letters of gold and are further marked on the neck as well as the body. p.s. The satin réclame series is only made in the first eight sizes. Refuse all bébés not having this guarantee".

To the left of the advertisement is a half section of a bébé's dress, showing the Bébé Jumeau band above the elbow on the puff sleeve. At the opposite side is a Jumeau chemise decked with lace and ribbon and displaying the Bébé Jumeau label on the waist band.

An artistic portrayal of the pampered French child was the ideal towards which Jumeau worked. This bébé, dressed in cream silk, sits amidst luxurious cushions. The bisque head is marked in red "Déposé Tête Jumeau" with the size "11". On the body is stamped "Bébé Jumeau Bte S.G.D.G Déposé". (Bébé Jumeau patented without government guarantee). She has fixed blue eyes, moulded ears and jointed wrists. The body is of the six ball jointed type. The original wig of real hair is still worn. Height 25 inches (63cms). Courtesy Emma Berry Collection. Photograph Acanthus.

A closed-mouth bébé with a blonde human hair wig over a cork pate. The head marked in red "Déposé Tête Jumeau Bte. S.G.D.G 11" with red tick marks. The doll has a closed mouth, fixed brown eyes and pierced ears. The jointed body carries a blue "Jumeau Medaille d'Or" stamp on the buttocks. It has unusual bisque hands. Wears the apparently original pale olive satin brocaded frock hung with tassels of pink and green beads and with panels of olive green velvet with matching velvet-lined straw bonnet. The bonnet is also decorated with olive satin and feathers. Height 23½ inches (60cms). Courtesy Sotheby's, London.

The line drawings in this advertisement give a general picture of the typical chemise and dress, as do the illustrations in contemporary catalogues, but it is much more difficult to establish the real detail of the heads and in some cases it is not even possible to see whether the mouth is closed or slightly open. A number of dolls marked "E.J." have been found wearing costumes in the style of the 1890's, so that it seems safe to assume that these traditional dolls continued to be produced alongside the open-mouthed bébés and the negro and mulatto heads. A few of the so-called Jumeaux Tristes have also been discovered in what appear to be the original costumes, which date from the 1890's and several of the coloured Parisiennes also date from this comparatively late period, when children were offered the full range of the firm's work. All the collector can do in establishing the date of manufacture is to examine the figure for the relevant clues. Are the shoes marked "Medaille d'Or" or simply "Paris Déposé"? Is the body marked "Diplôme d' Honneur"? Is it of the ball-jointed or the more realistic type with incorporated articulation? Is the box marked in any way? "Bébé Prodige", for instance, was registered in 1886, "Bébé Marcheur" in 1895 and "Bébé Française" in 1896. The box label might also furnish the date of the last event at which an award was won and Jumeau, being so acutely aware of his image, would not for long have continued to pack his dolls in outdated cartons, though it should of course be remembered that particular types continued in production after the amalgamation of

A closed-mouth Bébé Jumeau in completely original condition, with fixed brown eyes and the original mohair wig. The head is stamped in red "Déposé Tête Jumeau Breveté S.G.D.G" with the size "9", a red tick mark and the number "4". The later type body with jointed wrists bears the oval cream label "Bébé Jumeau Diplôme d'Honneur". The frock is made of ribbed watered silk and further decorated with velvet, edged with feather stitching. The pink satin shoes perfectly complement the frock, and are lined with white flannelette. They are marked on the leather soles "9. Bébé Jumeau Déposé". There are harmonising pink art-silk lace socks with simulated heel and toe sections. The original underwear is made from the pleated muslin that was so often used by the Jumeau dress designers. Height 21 inches (53cms).
Author's Collection. Photograph Acanthus.

several companies in 1899, so that a date can only be established when all the relevant information joins together to indicate a particular year. The majority of the dolls, often found without shoes, clothes or boxes, can only be approximately dated, and in many instances the year of manufacture concerns the collector much less than the quality of the head, making it necessary for each individual doll to be separately evaluated.

Mechanical and talking dolls

There is never any doubt regarding the date of the doll which had absorbed Jumeau's attention for so long; the Bébé Phonographe is immediately recognisable, as the torso incorporated a talking machine. By Christmas 1893, Jumeau was offering the public a size eleven doll with an open mouth, pierced ears and fixed eyes that stood 63 cms high, between 24 and 25 inches. The heads were marked "Déposé. Tête Jumeau. 11" and the bodies, "Bébé Jumeau Bte. S.G.D.G. Déposé" (Bébé Jumeau Patented. Without government guarantee. Deposited). An 1894 Bon Marché advertisement described the doll as "Bébé Jumeau Phonographe. Laughing, talking, singing, at will. Height 63 cms. Price 52 fr." Jumeau had designed this figure with great care and Lioret patented a spring that was small enough for a doll. The words which the bébé should utter were also only decided upon after long discussion, and Jumeau even organised a competition in "*Mon Journal*" in which children were invited to devise a suitable monologue.
The cylinder mechanism was concealed in the torso and key wound from the back. In the area of the chest was a removable metal plate for the insertion of the cylinders. A contemporary description explained that the sound was activated by a system of cog wheels that was set in motion by the pulling of the stop-start lever at the back. The doll then began to speak. "Je suis bien contente, maman m'a promis d'aller au théatre, je vais entendre chanter. Tra, la, la". The doll then goes on to sing a pretty song or even to laugh and ends by saying "Merci, ma petite maman".
The very characteristic cylinder with its red and pale green label bears around the circumference "Bébé Jumeau. Phonographe. Phonographe A.L. (a number) and Bte. S.G.D.G." Various cylinders in French, English and Spanish were supplied and in 1894 sold at four francs each. In accord with Jumeau's obsessive anti-German feeling, and, more practically, because few dolls were exported to Germany, this little child of Paris was not allowed to speak in that tongue.
Another cylinder plays "Bonjour, ma chère petite maman. Je suis bien sage et papa est très content. Nous irons voir Guignol pour l'entendre chanter. Pan! Pan! Qu'est-ce qu'est la? C'est Polichinelle, mamzelle. Pan! Pan! Qu'est-ce qu'est la? C'est Polichinelle, v'la. Au revoir ma chère petite maman. (Hello, my dear little mother. I'm very good and papa is very pleased. We are going to see Guignol and hear him sing. Knock! Knock! Who's there? It's Polichinelle, miss. Knock! Knock! Who's there? It's Polichinelle, that's who. Good-bye my dear little mother".
Many writers found the prattling of these French bébés irritating and unpleasant, and even today, the dolls, in comparison with their original production costs and the importance they were given by Emile Jumeau, are not as popular as might be expected. Their lack of appeal lies mainly in the fact that when they are costumed and displayed among other bébés, they look like the ordinary, basic, open-mouthed doll and examples sometimes sell at auction for much less than the dolls would seem to warrant. At the 1900 Exposition, a whole area was devoted to the Phonograph dolls and the public was

A Leopold Lambert automaton in which a Jumeau head has been used to create a most effective child's figure. The socket head has brown paperweight eyes, a closed mouth and pierced ears. The girl turns and nods her head as she raises and lowers the strings in each hand as though she is operating her marionette. The original silk costume is worn, decorated with lace, pearls and gold braid. The composition-headed Polichinelle was a favourite character in French children's stories and was almost always the mischievous doll in toy shop stories. Height 22 inches (56cms). Courtesy Sothby's, New York.

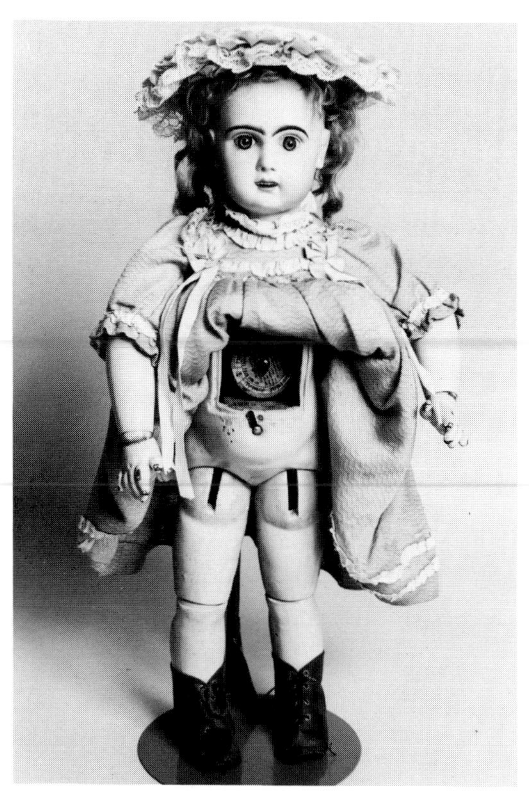

The Jumeau Phonograph, showing the complex mechanism contained in the torso. The phonograph dolls always have bodies of this late type in combination with open-mouthed heads.
Courtesy Christie's, South Kensington.

invited to watch the young women at work, recording the cylinders. One reporter obviously found the scene quite ridiculous with the girls seated at benches and "Laughing, crying and singing like lunatics". As the Phonograph dolls were quite expensive, the number sold was never very high, so that few have survived in fine condition, the majority having lost their cylinders, making their very average price even more remarkable.

Jumeau was also concerned with the manufacture of automata, though it is unlikely that he produced complete figures and certainly these are not mentioned in any of his advertising material or in the reports of his exhibits at the international exhibitions. Being an astute business man, he was aware of the fact that such complex figures were made much more economically by the specialist producers such as Decamps, and he was content to act only as a supplier of heads, usually of the standard types. Vichy, Lambert and Decamps all made extensive use of the firm's heads which are most frequently stamped "Déposé Tête Jumeau", many in fact belonging to the period after the amalgamation and exhibiting the high colour and poor quality bisque that would not have been tolerated by Emile. These early twentieth century automata were made as parlour amusements for adults and are often costumed either in some kind of fancy dress or in the style of an earlier period, making it necessary to date them mainly on the evidence of the textiles and the mechanisms. The problems connected with dating automata are compounded by the fact that so many have replacement heads, a popular dealers' trick being to replace a very basic German bisque head with another of the French closed mouth type, ensuring a much higher price.

In certain cases, where two dolls are included in a group, it is found that only one is made by Jumeau, sometimes both are obviously the firm's work but only one is marked, and there are even instances where the completely original dolls' heads come from both

An unusually pale open mouth Jumeau with fixed eyes, pierced ears and original long animal hair wig with its original mother-of-pearl slide. The shoes carry the bee mark and she wears the characteristic Jumeau printed cotton dress seen frequently in their advertisements. The later type body with jointed wrists carries the cream oval label reading "Jumeau Diplôme d'Honneur". The head is unmarked but incised "10" for the size. Height 24 inches (61cms). Author's collection. Photograph Acanthus

A Bisque-headed standing automaton figure of a smoker, 25½in. high, the head stamped in red "Dépose Tête Jumeau 3", the key marked "L. B."; French, late 19th century by Lambert, in wooden box.

France and Germany, the assembler simply choosing the faces he thought most suitable for the group he was creating. Some of the most beautiful automata are the closed mouth lady dolls who perform such actions as raising a basket lid to reveal a bird or even a small lamb sitting inside. Despite the fact that their appearance suggests an earlier date, they usually carry the red "Déposé Tête Jumeau" stamp. The child-like figures are more generally popular with doll collectors and also perform a variety of attractive actions to musical accompaniment, such as fanning themselves, opening an Easter egg, knitting or playing with a bird or toy. One particularly effective figure plays with a windmill, making the sails revolve, another pours tea and a small boy blows bubbles. The pages of Decamps' catalogues reveal the retrospection of many of these figures and, in a catalogue of around 1912, a girl is seen playing the piano and wearing the costume fashionable in the 1870's, while the smoking gentlemen, for instance, even when made as late as the 1920's, invariably wear eighteenth century style court dress. Considering the great skill of the house of Jumeau as a dolls' costumier, it is perhaps surprising that they could resist the temptation to complete the

When a child was presented with a Jumeau complete with its trunk filled with a great variety of attractive accessories, she must have enjoyed hours of quiet play. The bébé in the cream frock is an unmarked Jumeau with the head incised "4". She has an open mouth and a fully jointed body. Height 15 inches (38cms). The doll in blue with a closed-mouth and fixed blue eyes has fixed wrists. The head is incised "5" and the body is stamped "Jumeau Medaille d'Or Paris". She stands 14 inches (35cms).
The doll in cream courtesy Betty Woods. That in blue courtesy Lilian Middleton's Antique Doll Shop. Photograph Acanthus.

automata themselves, and their decision to leave the manufacture of such expensive items completely in the hands of a small group of specialist makers is a further indication of Emile Jumeau's sound commercial sense.

A few of the automata were intended for the amusement of rich children, the musical smoking monkeys, for instance, being thought of by D'Allemagne completely in the light of a child's toy rather than an adult amusement. The dependence of the French child on the diversions offered in her own home, even at the end of the nineteenth century when her American and English counterparts were enjoying a new found freedom, cannot be over-emphasised, and partly accounts for the very high standard of doll making.

Jumeau produced bébés and Parisiennes in the typically French manner for the girls he saw around him; he knew what they liked best and the degree of decorative detail that was demanded. It would not occur to any true Frenchman that the young of other countries should want anything different. Though the *jeune fille* of the 1890's had considerably more freedom than the child of the 1860's, she was still virtually a puppet in the hands of her parents, brought, exquisitely dressed, into the drawing-room to make a bow and speak a few polite words, for, above all, she was schooled in good manners. When she carried in her arms a beautiful bébé whose lovely face and elegantly contrived costume aroused the admiration of the company, the picture of the perfect daughter was complete.

Playthings for the perfect child

A feature of French children, that continued to perplex foreign visitors throughout the century, was the way in which they were always so anxious to make a good impression. Mortimer and Dorothy Menpes, writing in 1903 remarked that whether they were playing or skipping on the boulevards or at a fancy dress ball or dancing lesson, they were constantly aware of, and actually enjoyed the approving eyes of adults, so that a perpetual air of coquetry pervaded their smallest actions and it seemed as though they could not be happy unless they were admired. "No Parisian child will fly into a rage in public; to frown and stamp would spoil her appearance and ruffle her dignity. She would become terribly upset when walking in the boulevards if her dress were to become slightly soiled or buttons were to be lost from her shoe; she would hurry home by short cuts and unfrequented paths less perchance she should suffer the humiliation of meeting an aristocratic acquaintance."

Though the Menpes account is obviously based on wide generalisations, and concerns only the children of Paris, it does serve to indicate the atmosphere of absolute perfection that surrounded the French girl and made her quite different in needs and attitudes to those of other countries. Even when she was taken to the sands at Trouville, where a child would be expected to scamper and play, the authors watched diminutive figures setting their sashes in order, straightening sun-bonnets and neatening their curls, all with great seriousness and an eye for perfection.

It was for these cultivated children, whose upbringing was centred around the inculcation of good manners and an appreciation of fine things, that Jumeau produced dolls, from whose faces every trace of vulgarity or peasant ancestry had to be removed. The dolls were as placid and perfect as the parent wished the child to be, and to portray the ugly face of a child in a temper or sulking in a corner would have been completely

With the introduction of the open mouth, the general effect of the dolls became more childlike, though much of the detail of the earlier modelling was lost. This large example, dressed in a contemporary child's whitework frock and broderie anglaise cape, sits in a full-size chair. She has blue sleeping eyes, pierced ears and an open mouth with moulded teeth. The jointed later type body is unmarked but the head is incised "15" indicating the size. Height 32 inches (81cms).
Courtesy Lilian Middleton's Antique Doll Shop. Photograph Acanthus

foreign to Emile's thinking, and indeed to the demands of the French market. Though he frequently introduced Bébés with new titles, such as Bébé Moderne, Bébé Miracle and Bébé Prodige, they are all basically the same beautiful tranquil dolls and unless they are found in their original boxes with their marketing name, it is impossible to differentiate between them. Presumably Jumeau felt that at certain times the market demanded a new name and he obligingly revamped some of his stock. In much the same way, the so-called unmarked Jumeaux seem to have no significant difference of standard and we have to presume that the price differential was based mainly on the quality of the costume, the number of accessories and the quality of finish given to detail, such as the hair.

The only real source of information regarding the various differences in quality is found on the box-lid advertisements, such as one dating from 1897, which on the left side reads "National Toy. Bébé Jumeau with a wig of natural hair" establishing the fact that mohair was no longer the only wig making material. The same advertisement continues

> "The House of Jumeau. The most important maker in the world for the manufacture of beautiful bébés. The Bébé Jumeau is the national toy par excellence. It has appeared in all the major exhibitions, both in France and abroad, and has always achieved a brilliant success. The victories are important and in these commercial struggles they have propagated the renown of French manufacture in distant countries.
>
> Of extra superior manufacture, it is unique in its field and absolutely perfect, and if our charming girls have adopted it with joy and happiness it is because the dolls have indispensable qualities of perfection, beauty, strength and lightness.
>
> The Bébé Jumeau is beautiful. It owes this quality to the model made by the gracious master sculptor Carrier-Belleuse and its head, of fine porcelain, is ideally decorated, in a manner that distinguishes it from all others. The Bébé Jumeau is strong; it owes this strength to its method of construction. All the limbs are made of two different types of wood, morticed and glued together, which prevents them reacting to variations in atmosphere and prevents them from warping. The Bébé Jumeau is light, and although it necessitates a difficult task, this lightness is obtained by scooping away the inside of the limbs before joining the two pieces of wood necessary for their construction.
>
> With all these basic qualities the Bébé Jumeau can place itself in the first rank of French manufacture and can largely vindicate the name of the national toy".
>
> The label continues but in italics "The dolls are always placed in their boxes with their heads next to the side with the exterior label. Guarantee. The head of the Bébé Jumeau has been recognised as the exclusive property of the house by a judgement of the Court of Appeal in Paris. (A registered design mark). The form of the new body is the object of a registered design at the Conseil Des Prud-hommes.
>
> Patents
>
> The system for moving the eyes is patented, as well as the box which contains the doll. Counterfeiting is therefore illegal. The Bébé Jumeau is sold by all the leading shops.
>
> In 1881, 85,000 dolls made.
> In 1883, 115,000.
> In 1884, 220,000.
> In 1886, 300,000.
> In 1897, three million dolls' heads, white, black and mulatto".

The larger bébés were packed in particularly strong wooden sided boxes with cardboard

It is extremely rare to find a pair of dolls purchased by one family for its children and kept together to the present time. They were costumed in their original pinafores and whitework frocks by a kitchenmaid who worked for the Fitzroy family, whose laundry mark "V.F 14" was carefully embroidered on them. The heads, with replacement wigs, are very similar, having closed mouths and fixed blue eyes. The standing doll has applied ears and a body of the fixed wrist type. The head is stamped in red "Déposé Tête Jumeau Breveté SGDG 8" with a decorator's mark "N.o.2" and an incised "8". She is also marked on the body in pencil "Lady Fitzroy". Height 19 inches (48cms). In contrast, the seated doll has jointed wrists and moulded ears. Her head is stamped in red "Déposé Tête Jumeau" with a decorator's mark "18" and tick marks in black. The body is marked in pencil "Lord Chas. Fitzroy" and both are stamped in turquiose "Jumeau Medaille d'Or, Paris". The foreground doll stands 21 inches (53cms). Courtesy Lilian Wood. Photograph Acanthus.

An attractive bébé wearing a contemporary blue velvet coat trimmed with fur, a gingham frock and a lace-edged petticoat. Marked on head "C" above "E.J." for Emile Jumeau. Height 16½ inches (42cms).
Courtesy Sotheby King and Chasemore, Sussex.

lids and bottoms. The wood was attached to the cardboard with hand stitching for strength and the somewhat crude join neatened on the outside with strips of coloured paper. This construction ensured a rigid sided container in which the dolls could be shipped in safety and also meant that they were safe during the months of storage before the great October dispatch in preparation for Christmas.

One of the last authoritative accounts of a visit to the Jumeau factory was written by Leo Claretie, an enthusiastic chronicler of toys, who also supplied the report for the 1900 Exposition. Claretie is often an irritating commentator, as he was so concerned with his own emotional reaction to the toys that he often omits to give the sort of detail that came so automatically to the economist Rondot. This curious approach of the senses is sometimes particularly illuminating, as he does provide a picture of the smell and sounds of the busy workshops as well as a discussion of the product.

> "One passes through stores and shops where are piled up the basic material or half finished busts, arms, legs, eyes: it is a horror. One does not know in what obscure corner of the catacombs one finds oneself or if it is not the annex of the ancient charnel house of the innocents. The Innocents are here well named, for one would believe one saw an ossuary of new born children, greenish busts with holes for shoulders and groins, long lines of legs which resemble mouldering heaps of human hands grown still, piles of amputated arms, in an offensive odour of drying glue and old, rancid paper."

He describes in his book, "Jouets: Histoire et Fabrication", published in 1894, how he watches a boy twisting copper wire for the jointing hooks in the "Crochetterie" and how, in this hook maker's shop there was special machinery for making the spiral springs and hooks on whose strength the articulation of the bébé depended. This account

For the majority of collectors, the first Jumeau they are able to buy is one of the later open-mouthed examples, such as this girl with attractive pale colouring. Despite the advance of the open-mouthed doll with moulded teeth, traditional French refinements such as pierced ears and cork pates continued to be used. This doll is of the later type with jointed wrists and weighted eyes. The bisque head is marked only with an incised "X" and the size "6". A contemporary whitework frock is worn. Height 18 inches (46cms). Courtesy Emma Berry Collection. Photograph Acanthus.

indicated several advances since the publication of the "Notice", such as the use of two-part heavy steel matrices so that the torsoes could be moulded in one piece for extra strength. During the manufacture of this part of the body, which had to stand a great deal of strain, the workmen filled the mould with glue-soaked paper, pushing the substance into every hollow to ensure an even inner surface. The torsoes, as in the earlier dolls, were made even more rigid by the insertion of a wooden bar to which the copper linking devices and the elastic were attached. In some of the stomachs, small whistle-like voice boxes were added, "One of the most ingenious types with springs, a box for air valves, a little apparatus for words."

The bébés' limbs, moulded in other steel matrices, were also at first a dirty greenish colour, though the collector should bear in mind the fact that the mixture would have varied according to the waste materials used and was probably just as frequently grey. There were advances too in the blacksmith's shop, brought about by the introduction of the counterweight eyes and on the ground in front of a shearing machine were dozens of small triangular pieces, the supports for the balls of lead to which were attached the opening eyes.

In general, the process of manufacturing the bébé had not changed greatly since 1885, and the repeated straining of the paste through the finest mesh before the manufacture of the heads began was again stressed. The most significant difference was with regard to the moulding, as this was now effected by pouring the slip into the mould, allowing it to stand for a few minutes, then pouring away the surplus. In his account of doll manufacture provided for the 1900 Exposition, Claretie adds that the pouring process was by means of a tap. When still in a malleable state, the eye sockets were cut out, at this stage it would seem, almost invariably by the use of a template as the heads of this, the last decade of Jumeau's reign, rarely have that slight unevenness which gave added appeal to the fixed eyed bébés of the 1880's. Presumably the device designed by Jumeau for the complex eye-lever mechanism was gradually put into use for the majority of heads, especially as the counterweighted sleeping-eye mechanism also depended on regular eye sockets.

Claretie also disliked the sleeping eye mechanisms, which he felt were completely unrealistic and he particularly hated the scaley effect of the paint on the lids. He described the use of the four directional eye mechanism moved from right to left and up and down by means of a small knob at the back of the head, a method that can have had little lasting popularity, as so few examples are now discovered and it seems likely that the child found the broad inverted "T" shape which the movement necessitated, too unrealistic. The method of fixing the eyes varied in accord with the price of the doll; there were static eyes, moving eyes, living eyes, automatic eyes and several others. "May I say that these movements of the eyes are horrible to see because they are atrociously imitated, in spite of recent progress". Previously, the top section of the counterweight eye was painted in flesh colour but this soon became scratched so a new method was devised in which the colour was fired into the enamel and was unalterable.

In the studio, where the heads were painted, great care was needed for the drawing of the eyebrows in neat, parallel brush strokes and he comments that these were not as well executed when the worker was troubled, had a headache or had gone out the night before. In his Exposition report he added that Monday morning heads were always particularly bad.

Another difference from the period of the "Notice" is that by the mid 1890's, it would appear from Claretie's comments, all the costuming was executed at the factory itself. The shoemaker's shop, at the time of the visit, was producing uppers that were cut of

An unmarked Bébé Jumeau (seen also on the dust jacket) with a hand knotted fair mohair wig and pierced ears. The head has red and black tick marks and a painted "7" in red for the decorator. The body is of the later child-like type, with jointed wrists. She wears the original black leather shoes marked "Paris Déposé" with the bee mark. Size 9. They are decorated with dark red rosettes. The reverse side of the toy watch can be seen in this photograph. She carries a bisque headed German folie with a squeaker and bells. Height 20½ins. (52cms.). Author's Collection. Photograph Ancanthus.

Though catalogues mention "Oriental Jumeau", what was usually produced was the standard bébé dressed in Eastern costume. This late nineteenth century doll was beautifully costumed and supplied with a specially shaped Oriental wig that appears to have come from the factory. Such a complex embroidered costume must have made the cost of the doll quite high. The doll has the later type body, with jointed wrists, an open mouth, pierced ears and fixed eyes.
Courtesy Worthing Museum and Art Gallery.

bronze kid, glued to an inner sole and then covered by a realistic outer surface applied by another worker. The sewn lines of a full sized shoe were imitated at the factory by machine stamping. The basic footwear was then decorated with rosettes, ribbons, cockades and imitation buttons. A whole variety of tasteful shoes were created including pumps, dancing slippers of red and white satin and tiny ladies' boots. From the walls hung the designs from which the shoemakers were to work and these also gave size guidance, some of the patterns being large enough for a six year old child.
The walls of the costuming room were decorated with models of dresses, displayed between the zinc patterns and the boxes of sewing machines. "Women are busy titivating these little ladies, combing their hair, dividing the fringe on the forehead, pulling on the stockings, adjusting belts, arranging necklaces, putting in earrings and assuring the new Galatea of her status as a Parisienne, which is like her mark of origin." Curiously, much of the costuming, according to Claretie, was carried out before the head was attached, though prints of the factory do not confirm this. The whole of the costuming process was under the control of Madame Jumeau, and Emile claimed that it was her good taste that was exhibited.
In the wigmaker's shop another advance was discovered, as real hair instead of mohair was used for the most expensive bébés. Bales of Tibetan goat hair were stored at the factory and the small strands drawn out, rolled on wooden pins and wrapped around with paper. They were then boiled and dried out in gas fired cupboards. This process made some of the coarser hair somewhat stiff in appearance but it did mean that they stayed firmly in place. These little curls were then sewn in neat rows around the fabric

Seated in an Edwardian wickerwork armchair is this small doll from the late nineteenth century. The body is of the late "improved" type that aimed at a more realistic interpretation of a child's shape. The bisque head is incised with the size "6" and is marked in red "Déposé Tête Jumeau". She has pierced ears, fixed brown eyes and a closed mouth. The original shoes and socks are worn, the shoes carrying the "Paris. Déposé" mark with a bee. A contemporary outfit of coat and frock is worn. Though probably commercial, it was not supplied by Jumeau but possibly bought especially for the doll. Height 16 inches (41cms).
Courtesy Emma Berry Collection. Photograph Acanthus.

skull caps, with curls in front and plaits behind, before the completed wig was nailed to the cork pates. Doll's hair was one of the most expensive raw materials needed and all the makers controlled its use very carefully, so that the extravagance of many of the Jumeau wigs, with long tresses and a generous use of the mohair or human hair, made their products stand out from those of their German competitors.

Influential patrons, visiting the firm were presented with statuettes, modelled by the artists in the studio that was still the centre of the factory. Presumably these figures were not marked, as none are known to have survived, but they served to emphasise Jumeau's ascendancy over all other dollmakers, who had to rely on ceramic work carried out by others. Claretie first entered the sculpture and moulding rooms where "Artists chisel in plaster faces of all sizes and expressions", which were later cast to form moulds. As the marked Jumeau heads of the 1890's are so similar, it is difficult to imagine why the artists were creating faces that appeared so different in expression; was this yet another of Claretie's exaggerations, or is it possible that Jumeau produced some heads for other companies?

In his 1900 Exposition report, Claretie adds a vivid description of the constant noise encountered in the factory, a combination of the bubbling of the composition mixtures in the cauldrons and the grumbling of the machines that were attended by sweating, half-naked men, who, one might think, were producing naval equipment rather than dolls. "The little doll in the girl's arms could not wish to return to this birthplace, where heavy weights crashed noisily into moulds when its body was formed of sickly smelling paste." The 1894 description of the manufacture of the separately articulated hands sounds equally unattractive as "All day long the piston comes and goes, with each stroke it drops a hand, a little veined hand, with nerves and articulation with bones, glistening with the oil of the machine and of a nasty brown colour. The little hands fill baskets, are dried at the end of thin sticks which are thrust into planks, one would think to be crossing a terrible place of plaintiffs where hundreds of pygmies have passed under the axe of the great executioner."

Character Heads

Claretie's description of artists working in plaster on faces of all expressions is one of the most tantalizing as it is always so difficult to distinguish between his usual poetic exaggeration and fact. Does the comment perhaps prove that work had already begun on some character type heads, or does it simply refer to some variety made especially for automata or special figures?

The first section of the series of character heads produced by S.F.B.J. has always aroused controversy among researchers. Auction rooms and collectors date the few examples that appear very optimistically, ignoring the facts that in mood, quality of bisque and effect, these heads are completely in the twentieth century play doll idiom. A few examples have appeared mounted on Medaille D'Or bodies but, as every researcher is aware, once a piece has passed through the hands of a dealer it is hard to be sure of complete authenticity. During the last few years we have all noticed the number of pouting Heubachs, for instance, originally sold on bent limb bodies, that have been transformed into the much more expensive double jointed toddlers. Without a much larger sample it is impossible to be sure of the original presentation of these dolls.

One of the controversial Jumeau characters. Few examples of these interesting dolls are found, so it is difficult to compare body markings to be sure of authenticity. This exceptionally fine character has half-closed, fixed glass eyes, and an open-closed mouth with six upper and four lower teeth and the effect of a tongue. She wears the original cork pate and has a black mohair wig. This body is stamped "Bébé Jumeau Hors Concours 1889 Déposé". The head is stamped "Dépose Tête Jumeau Bte. S.G.D.G." in red and is incised with the character series number "208". It also carries tick marks. Height 22 inches (56cms). Courtesy Sotheby's London. Photograph Sotheby's.

Typical of the skill of French dressmakers are these dolls' costumes illustrated in the catalogue of Magasins de la Ville de Saint Denis in 1902. Costume of dolls seems to have changed little in style between 1890 and 1905.
Courtesy Justin Knowles.

I would not find it too difficult to believe that these heads were designed for automata in the nineteenth century, as the type and variety of models used for these always varied from those intended for dolls. I find them very hard to equate with Emile Jumeau's general production, as they were simply too much in advance of their time. It also seems unlikely that no reporter, commentator or judge remarked upon models that would have been truly eye-catching at this period. When the German Kämmer & Reinhardt characters appeared in the early twentieth century, they attracted very widespread interest. Surely such exaggerated dolls' heads appearing among the delicate pretty children would have aroused some comment.
The marking on the few recorded early heads is also strange comprising a large "2", a small "o" followed by a large number, a method quite different to S.F.B.J. marking of characters, where the numbers are the same size. In a most fascinating piece in the American "Doll News" (Summer 1980), Ralph Griffith showed a series of photographs of extremely rare heads of this type that were in his own possession. He was also unsure when they were first made. He showed seven character heads, all of which were marked

The early portrait Jumeau shown in close-up to illustrate the way in which the cutting of the eye sockets varied according to the individual worker. Her eyes are of a slightly different shape, and it is noticeable that the pupils are also of a different size, all underlining the early date, before mechanical methods of standardisation were used. She has applied ears and a hand knotted mohair wig. It is unusal for dolls of this fairly small size to have applied ears. She wears a contemporary straw flower-decorated hat. Height 18½ inches (47cms). Author's Collection. Photograph Acanthus.

with the "Déposé Tête Jumeau" stamp, a method that continued to be used long after the amalgamation. The "2 o 1" has an open-closed mouth with moulded teeth, the "2 o 3" smiles in a more restrained way. In a sad faced example the number "206" is incised actually inside the head. The most remarkable of all character dolls, a girl with her moulded tongue pushed right out, again has the number "217" incised within the head, a doll made, if possible, even more desirable as it has applied ears.
The methods of numbering and the stamping all indicate a twentieth century date, but until more examples on original bodies appear we cannot be sure that any of these were produced commercially before the amalgamation, especially as they are so much out of tune with the general mood of Emile Jumeau's work. From Claretie's comment on faces of "all expressions" it is tempting to think that several of these models were at least designed before the amalgamation, though it is unlikely that current taste would have enabled them to be marketed as play dolls. Hopefully as more completely original examples appear we can come to more definite conclusions regarding these most tantalizing and highly collectable characters.

The end of the era

Claretie's earlier account describes a factory that was hard at work producing good quality dolls in ever increasing quantity and reveals nothing of the great pressure of German competition. In France, all the production of dolls was in the hands of skilled adults, work of this type being considered too demanding for the use of child labour, which was never as extensively used as in Germany and Britain. Wages were consequently comparatively high, whereas the German toymakers, often working in country districts, made use of the cheap peasant and child workers, whose time could be purchased inexpensively. The German dolls were also much more economically produced, from cheaper materials and by the use of simpler methods of articulation. It was unfortunate, from the French point of view, that when costumed, these much cheaper dolls were extremely attractive and combined the added allure of a much lower price ticket. Little wonder that Emile Jumeau was forced to offer his bébés at special rates. While dolls could be relied upon to sell because of the beauty of their costumes, the French doll could more than hold its own but, as children's clothes became much simpler, the doll no longer needed such exotic garments and those dressed simply probably appealed just as much to both parents and children. Jumeau must have realised that it was impossible to produce quality dolls at a lower price and was eventually forced to consider amalgamation with other French makers as the only real hope of survival. By forming an association, it should have been possible to improve the merchandising of the dolls, always weak in comparison with that of the German makers, and to arrange displays of dolls at exhibitions and trade fairs, as well as publishing catalogues and indulging in more widespread advertising. After a difficult period for trade in the mid 1890's, the Société Française de Fabrication de Bébés et Jouets was formed in 1899. The Official Municipal Bulletin of April 15 declared that the Society had acquired all the important makers of dolls and were assured of the industrial and experimental competitiveness of the syndicate. S.F.B.J. consisted of ten associates and its registered office was at Jumeau's address, 8 Rue Pastourelle, Paris. Its capital was 3,800,000 fr. In uniting the making of the different makes of dolls in the

Sailor suits were as popular in 19th century France as in Britain, and Jumeaux are found in both blue and red versions made from the traditional fine serge. This bébé carries the red "Déposé Tête Jumeau" stamp on her head together with the size "10". She has the later type body with jointed wrists which is unmarked. Her head is particularly nice, as the quality of the bisque and the colouring is good and she has the advantage of the desirable applied ears. She has a closed mouth, fixed brown eyes and the original hair wig. Height 22 inches (56cms).
Courtesy Lilian Middleton's Antique Doll Shop. Photograph Acanthus

same factory it was hoped that several goals could be achieved. The Bulletin set out these aims.

"1. To produce work in conditions that are economic and by this fact to make great progress in the making of dolls and toys.

2. To achieve a reputation for our products in all the world markets by means of prices which are relatively advantageous.

3. To enable the French clientèle to have all the advantages which they have the right to expect."

The various members were then listed, together with the trade names that they brought to the company. Among these Jumeau was the most important. "Emile Jumeau and Madame Ernestine Ducruix, his wife, living together at 67 Boulevarde Beaumarchais bring to the company the funds of the trade of the making and selling of dolls, 8 Rue Pastourelle, Paris and lease to the company for ten years the two factories situated at Montreuil sous Bois, one at 152 Rue de Paris, the other at 64 Rue François-Arago. They bring notably the mark Bébé Jumeau."

It is sometimes suggested that Jumeau continued to exercise great control over the dolls that carried his mark after the 1899 amalgamation, but the Bulletin states that "All the associates declare that they have ceded to S.F.B.J. all their trade marks, makers' marks, all patents and registered designs which are dependent on them as well as their materials, machines, tools, agencies and objects of furniture concerned with exporting and selling. The company has nine factories, the manufacturing moulds of Bru, Gaultier and Fleishman & Bloedel will continue to be used."

It is obvious from the quality of a number of dolls marked Jumeau and made before the First World War that the heads were still a source of pride to the workmen who produced them and the decline in quality was therefore only gradual. The costume of the dolls was still of a good quality and attention was given to the smaller details such as the hair and the stockings so that it is often difficult to be sure whether a nice quality head was made in 1897 or 1910. Unfortunately the completely French origin of the Bébé Jumeau was not maintained, and the syndicate saw fit to purchase the much cheaper German-made heads which could be mounted on Jumeau marked bodies and sold in Bébé Jumeau boxes. To the purist, these dolls of the amalgamation cannot be thought of as true Jumeaux: the finest are only those made under the personal supervision of Emile and Ernestine.

At the 1900 Paris Exposition, Jumeau bébés were displayed by S.F.B.J. "Elegant and haughty with their silken dresses, the frills of lace, the frou-frou of their rich blouses and the voluminous chic of their feathered hats, a graceful effect which is achieved by the use of delicate shades of pink and blue." Before the formation of the syndicate, it was only Emile Jumeau's factory that had the equipment and the kilns for firing bisque heads, but the company had recently installed a much larger kiln where thousands of heads could be fired at the same time. All the machinery, the crushers and mixers were driven by a 15 horse power engine and it was the only installation of its kind in France for the manufacture of dolls.

It was possibly the prospect of such dazzling improvements that had encouraged Jumeau to form part of this syndicate, which was described in the Exposition report "There is no longer a Bébé Bru, Bébé Géant, Eden Bébé or Bébé Jumeau; there is only the Bébé Française, made in concert by all the associated firms, which partake of the dividends of the operation. The different dolls continue to be born in their own factories. They all go to be dressed in the central fashion area of the Picpus, that old quarter of Paris that seems to be in the country." Presumably some of the Jumeau bébés were also sent out to be dressed after 1900. Little wonder that the quality of the costume gradually declined after the amalgamation.

The question of why Emile Jumeau, at the height of his manufacturing success, decided to abandon his individualism will continue to puzzle. The weight of German competition was a force that he had contended with since he took over the factory and the whole of his life had been directed towards manufacturing the supreme French doll, embodied in the name of Jumeau. Why was all this endeavour thrown aside and the Empire of which he was so proud combined with those of his rivals?

Perhaps much of Jumeau's advertising which had painted such a glowing portrait of the factory, was founded on wishful thinking, perhaps he was, in his later years, able to see that an individual company could not hope to compete with the cheap and effective German dolls. That Jumeau still dreamed of an Empire that stretched across the world is revealed in his 1890's advertisements and the information he provided for visitors to his factory, though a note of chill was struck in 1895 when he was forced to offer a third quality doll with a reduction in price of as much as 60 per cent, an offer that could only have been made in a desperate attempt to increase turnover. Curiously, little is known about this very significant period and his file at the Legion of Honour reveals little, not even the date of his death, apparently in 1910. Though Emile Jumeau's achievements were very great in the development of the French doll, they were obviously lightly regarded in the context of contemporary French history and Emile's file is one of the smallest they possess. Possibly this lack of national recognition for his work dispirited him, for he was to be awarded no further public honours despite his constant success at International Exhibitions and his acknowledgement seems to have been kept very firmly in the area of dollmaking. In 1897, Emile Jumeau was fifty four, yet, within a few years he and Madame Jumeau had retired, allowing his two daughters to be involved with the syndicate but remaining for a while as directors. It seems most likely that, having only daughters to carry on the factory, Emile was finding the strain of international competition too great: perhaps he was ill. He at least made certain that the name of Bébé Jumeau should continue.

With the amalgamation, the period of the first quality bébé ends, and within a few years the "Déposé Tête Jumeau" stamp was to be seen on dolls of such inferior quality that Emile would not have acknowledged them. The syndicate's use of German made heads on French bodies must have appalled Jumeau, whose entire business life had been spent in fighting against their use. It must have grieved him to see German heads used, purely for the sake of economy, on the bodies he had spent so long in perfecting. Their use was, in fact, a complete return to the first half of the nineteenth century, when German porcelain and carton heads were used on French leather bodies. It was the negation of his life's work. Or was he perhaps just an advertising showman, patriotic when it suited his ends but quite willing to abandon his ideals for a more commercial proposition? Patriot and philanthropist or shrewd opportunist, taking advantage of the nationalistic fervour of the French nation after the Franco Prussian war? His true portrait is obscured by his self-advertisement but the bébés remain, beautiful, serene and unmistakeably French. Perhaps they alone were the true patriots.

Atelier des friseuses.

Appendix 1

Patent 177127 dated July 1st, 1886

"In the heads of dolls and others for opening or closing of the eyes by the rotation of the eyes themselves.
They are characterised by the mounting of the eyes on a swivel rod, by the method of rotation of the said rod, which is fitted with eyes on metal bearings which are set back and sealed and by means of fashioning the opening of the eyes in the mask in such a way that the eyes have a regular movement as long as they are hemispherical so that their movement can take place without leaving any gap around the eyelids.
Further, the method of construction, entirely metal and mechanical used for the fitting of the different constituent parts of this improved mechanism, ensures that it will be very solid and will be able to be made at an exceptionally low price.
This improved mechanism is shown in the drawing which shows both the assembly and the function.
Figure 1 Back view. The inside of a mask of a doll, fitted with my mechanism for moving the eyes.
Figure 2 Vertical cross section of the profile.
Figure 3 Horizontal cross section done at the level of the eyes.
Figure 4 One of the eyes used.
Figure 5 The mounting of the eyes.
Figure 6 One of the bearings.

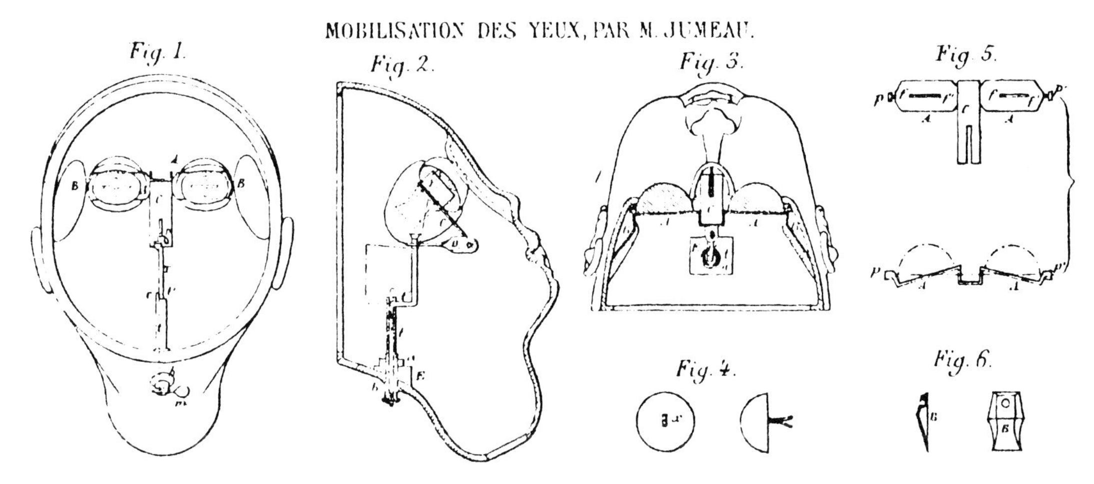

In the head shown the eyes are of a hemispherical form and have a metal tail (letter X Figure 4), allowing it to be fixed to the transverse metallic member (A), of which the extremities swivel with the minimum of friction in the bearings (B figure 6) set back and sealed onto the mask by the use of some adhesive mastic or other.
The tail "X" is introduced into a horizontal slot f f′, of the transverse (Figure 5) and then bent back in the manner of Parisian buckles and the whole is fixed by means of a little mastic after care has been taken to ensure that the eyes are set regularly in the orbits.
The transverse A, or eye movement is bent in such a manner as to give to the eyes the horizontal direction desired and the extremities are bent as in figures 3 and 5, to the geometric line which joins the pivots or swivels p,p′ which passes exactly through the centre of the ocular globes. This last condition is indispensable for hemispherical eyes turn exactly around their centre.
So that the eyes may move in proper conditions it is also necessary that the openings which receive them are made in an absolutely mechanical fashion. For this reason I have made a mould of the mask in the appropriate material which the workman applies to the dolls head to allow the mechanism in question to trace on it with a point the outline of the openings which he will then be able to cut with complete security.
In this manner the spacing of the eyes for the same model of head is constant and the section of the opening allows the eye, in combination with its hemispherical form and its movement of rotation about its centre always to be in perfect contact with the visible edge of the eye socket, which would be obtained with difficulty, if ever, with the current procedure, where all is left to the capability and skill of the workman.

In all cases my special procedure, by use of a mould, eliminates all trial and error and allows one to obtain a good result at the first attempt. Also the bearings (B), advantageously replace the current means of turning the eyes within a shell of plaster. That said, I will indicate how the mounting of the eyes is activated. This mounting or transverse A is joined by a bar, C with two branches which, when it is stamped, is bent back on itself as in Figure 2, to grip a swivel to which is attached a metal connecting rod, D, then bent in the manner shown in the perpendicular view. The lower eye of this connecting rod receives the end of a metal stem T, which runs through a tube, "t" in which it is a very tight fit, then this tube "t" is fixed to the mask by a nut "a" which is engaged on a little block of wood, E. The tube "t" is threaded for this reason and has an embase b. The stem T comes out behind the neck and is fitted with a little metal knob which suffices for turning it to the right or left to produce the effect desired, that is to say, opening or closing the eyes. The amplitude of the movement is limited by a prolongation c of the tube which acts as a stop for the bent over section "t" of the stem T figure 3.
The easy movement of the eyes is effected by a piece of leather glued to the inside of the head around the eye socket.
Finally I would observe that the majority of the pieces of this improved mechanism are obtained economically by mechanical cutting and that the said cutting is followed by the stamping of the transverse A and the bearings B to give them the shaping necessary for their function. N.B. The form and dimensions of all these parts are relative, their sizes notably depend absolutely on the size of the head to which my mechanism is applied".

Appendix 2

Patent 182307 dated March 21st, 1887

"In the heads of dolls or others use is made of moving eyelids which cover fixed eyes to simulate sleep.
With moving eyes the eyelashes stay fixed, with moving eyelids the lashes are lowered. In the first case the eyes are in the right place but the illusion is incomplete. In the second case the illusion is complete but the eyes are too deep set, which detracts from the good effect of the head.
To remedy these diverse defects I thought of moving the lashes with the eyes and not by means of any eyelid whatsoever, for this reason I have had recourse to the method which is the object of the present invention.
The eyes are still of a hemispherical form and are fixed on a transverse metallic bar, A figure 1, whose ends swivel in bearings which are sealed to the interior of the mask.
A mechanism gives to the eyes a sideways movement in regard to the orbits or openings made in the mask (the mechanism shown in figure 1 is that described in my patent of July 1, 1886; the reference letters used there have been kept.)
The eyes of pure enamel made by the ordinary means receive the following preparation before being fixed on the transverse A.
A little above the iris, that is to say in the area where the eyelid ends when the eye is open, a line is traced right the way round and the globe is cut in two parts (M, N, figure 2). Between these two parts the lashes Y are inserted and the parts are glued one to another with suitable mixture and the upper part is given the colour of the mask as it will represent the eyelid.
This tinting can be done before or after gluing and all normal methods can be used to achieve this.
Instead of cutting the eye-ball after making it, the separate parts, m, n, can be obtained separately adjusted by use of a grind stone or in some other way so that they fit each other well and then insert the lashes before gluing. All other means of inserting the lashes into the eyes can be employed.
Whatever the means used, the eyes thus supplied with lashes are then fixed on the transverse A by means of a metal tail, x, or some other means and the whole is mounted into the mask by normal methods.
When mounted in a head in this manner, when the eyes are open the inset lashes y, touch the upper edge of the orbit where they stand out in relief and give great charm to the face. To complete the illusion, one can glue lashes, z, to the lower edge of the orbit.
When the eyes are closed in the head by moving the mechanism the upper lashes, Y, are lowered by the same movement and come down to touch the lower lashes z. The part n, then appears and gives the illusion of a real eyelid, taking with it the upper lashes.

The moving of the lashes by the eyes themselves constitutes a real industrial advance; they can be applied economically to expensive or more common toys to which they assure a considerable advantage, both from the point of view of the complete illusion and from the good effect of the eyes which will always be in just the right place and not too set back as is the case with moving eyelids".

Appendix 3

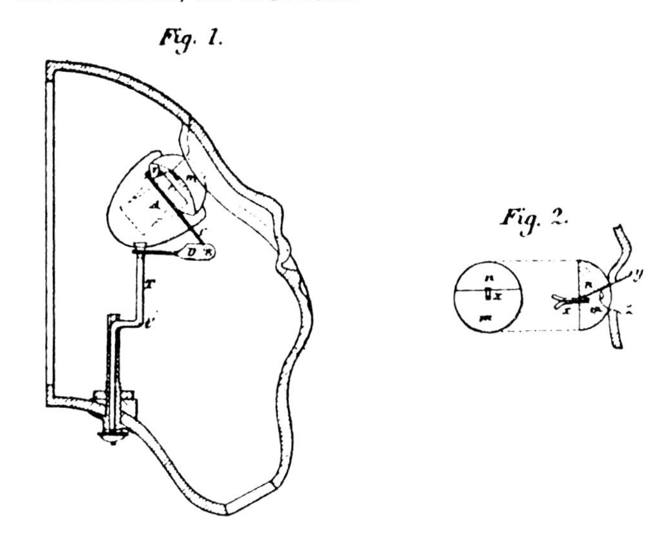

The drawing accompanied the patent specification to show, in particular, the angle at which the lashes were set.

Schedule of Prices and Sizes

Source														
Grand Magasins du Printemps 1887 (Price in Francs)		26cm	29cm	32cm	35cm	38cm	42cm	48cm	54cm	60cm	70cm			
With Chemise		3.50	4.50	5.50	6.90	8.50	9.50	12.50	17.50	21.50	30.00			
With Silk Chemise		4.50	5.50	6.50	8.50	9.50	11.50	13.50	18.50	24.50	34.50			
With Silk Dress		8.50	12.50	14.50	19.50	21.00	24.00	29.00	—	—	—			
With Rich Silk Dress		11.50	14.50	17.50	21.00	26.00	31.00	37.00	52.00	63.00	—			
Size Numbers	1	2	3	4	5	6	7	8	9	10	11	12	13	14
Bon Marché 1894 (Price in Francs)	25cm	28cm	31cm	34cm	38cm	41cm	45cm	49cm	54cm	58cm	63cm	68cm		
Closed Mouth					7.50	8.50	10.50	12.50	15.50	18.50	22.00	27.00		
Open Mouth					—	—	13.50	16.50	19.50	22.00	27.00	32.00		
Phonographe					—	—	—	—	—	—	52.00			
Maison de Petit Saint Thomas 1895 (Price in Francs)														
Dressed	9.50	12.50	15.50	19.50	23.00	27.00	33.00	39.00	45.00	55.00	65.00	75.00		
Naked	3.25	4.25	5.25	6.50	7.50	8.50	10.50	12.50	15.50	18.50	22.50	27.00		
Talking (Unsized)	—	—	—	12.50	15.50	19.50	21.00	—	—	—	—	—		
Hamley's Late 19th c. (Price in Shillings & Pence			31cm	33cm	34cm	38cm	43cm	46cm	49cm	57cm	61cm	66cm	70cm	76cm
First Quality			—	—	—	—	14/6	17/6	20/6	25/6	29/6	36/6	42/6	54/6
Talking			—	—	—	—	19/6	—	26/6	—	37/6	—	—	—
Second Quality			5/9	6/3	7/9	9/6	10/6	12/6	14/6	17/9	21/6	—	—	—
Talking & Moving Eyes			—	—	—	—	15/9	—	21/6	—	29/6	—	—	—
Heads Only (Inc. Fixing)	2/-	2/3	2/6	3/-	3/6	4/-	4/6	5/-	5/6					14/6
Marshall Field (Price per Doz. in Dollars)														
In Chemise	18.10	22.50	27.00	33.00	39.00	42.00	48.00	54.00						

NOTE: Size numbers relate to Bon Marché, St. Thomas, Hamley's and Marshall Field. It is interesting to note that the centimetre size differs between stores.

JUMEAU MARKS

Marks on Heads

J Incised	DEPOSE E C J Incised	DEPOSE E 8 J Incised	8 E J Incised
JUMEAU Incised	JUMEAU 1 R Incised	DEPOSE TÊTE JUMEAU Bté S.G.D.G. 10 Stamp	TETE JUMEAU Bte S.G.D.G. Stamp
DÉPOSÉ TÊTE JUMEAU 9 V L V Stamp	DEPOSE JUMEAU 7 Incised	BÉBÉ FRANÇAIS	B.12.L Incised

Body and Shoe Marks

JUMEAU MEDAILLE D'OR PARIS Body	JUMEAU MEDAILLE D'OR PARIS Body	BÉBÉ JUMEAU BTE S.G.D.G - DÉPOSÉ Body	BÉBÉ JUMEAU DIPLÔME d'HONNEUR Body
BÉBÉ JUMEAU *Diplôme d'Honneur* Body	E. JUMEAU MED. OR 1878 PARIS Shoe	9 PARIS DÉPOSÉ Shoe	10 BÉBÉ JUMEAU MED. OR 1878 PARIS DÉPOSÉ Shoe

Chronology

1842 Pierre François Jumeau listed in Paris Almanac in partnership with Belton.
1843 Jury of Vienna Exhibition of 1873 informed by Jumeau that the company began manufacture in 1843. (The Jumeau "Notice" also gives this date.)
1844 Jumeau and Belton awarded an Honourable Mention at the Paris Exposition of Industry. The address given as Rue Salle-au-Compte.14. They displayed "naked and dressed dolls."
1848 First mention of address as 18 Rue Mauconseil.
1849 Bronze Medal awarded at Paris Exposition. At this time making leather-bodied dolls, some with carton and others with wax heads (from England). 65 women were employed and the turnover was 120,000 francs.
1851 Great Exhibition, London. Firm awarded a Council Medal for excellence in costuming dolls. Jumeau was the only Parisian maker to exhibit dolls.
1855 Rondot, in a report on the 1851 London Exhibition describes Jumeau as the most important dollmaker in France.
1855 Universal Exhibition, Paris. P.F. Jumeau awarded a 2nd class medal. The term "bébé" first used at this event to describe articulated dolls. His senior factory worker also awarded a 2nd class medal for the costuming of dolls.
1859 The term "bébé" first used in the Paris Almanac of Commerce. The Paris Commercial Dictionary commented that porcelain was used by Jumeau.
1861 Paris Almanac of Commerce mentions talking dolls.
1863 "La Poupée" showed patterns for lady dolls made by Jumeau.
1867 Silver Medal awarded at Paris Exposition. The Jumeau address given as 8 Rue d'Anjou au Marais. He was described as one of the largest makers. The dolls had porcelain heads and skin bodies. Some had swivel necks, others of the cheaper type were fixed. Both painted and glass eyes were used. The wigs were of mohair or fur.
1873 A larger factory set up at Montreuil sous Bois.
1873 Vienna Exhibition. A Medal of Progress and a Gold Medal awarded. His son, Emile Jumeau was mentioned for co-operation in the production of dolls in this award.
1876 Philadelphia Exhibition U.S.A. A Gold Medal for fashionably dressed dolls.
1878 Emile Jumeau in control of company and described as a maker of kid and jointed wooden dolls in the Paris City Directory.
1878 Paris Universal Exhibition. A Gold Medal. The Medaille d'Or stamp relates to this award.
1879 The Paris City Directory of 1885 mentions "unbreakable bébés" in Jumeau's listing and describes them as "unique models" introduced in 1879.
1880 Melbourne Exhibition, Australia. A Gold Medal.
1882 The Paris City Directory stated that in 1881 some 85,000 dolls were sold by Jumeau and that sizes 9-16 had paperweight eyes and applied ears. Jumeau himself did not recommend his old style leather-bodied dolls, which were now produced only to order.
1884 New Orleans Exhibition. A Gold Medal awarded.
1884 Talking dolls claimed to be a new creation.
1884 The annual turnover claimed by Emile to have risen to half a million francs.
1885 The Paris City Directory comments that all the dolls carry the Jumeau name.
1885 Emile Jumeau made a Chevalier of the Legion of Honour.
1885 A Jumeau advertisement describes the "Bébé Parlant" made in sizes 7-12. A new series of bébés also marketed with a green arm-band in sizes 1-6.
1885 The Jumeau "Notice" mentions that the manufacture of glass eyes for the dolls was at the Rue Fontaine au Roi. Some of the bébés had a simple speaking mechanism worked by a pull-string.

CHRONOLOGY

1885 Jumeau claims to have 500 employees.
1885 Antwerp Universal Exhibition. Diploma of Honour awarded. The first time this high honour had been awarded to a toymaker. This mark used on the majority of dolls after this date.
1886 A Patent for moving eyes (No.177127).
1886 Bébé Prodige registered.
1886 Mark No.24407 registered for Bébé Jumeau. Gold on satin.
1887 Advertisement claimed that more than 130,000 bébés were sold.
1887 Patent No.182307 for sleeping eyes.
1888 Trademark No.24407 also registered in U.S.A.
1888 First reference to Jumeaux with open mouth and teeth.
1888 Article in "La Nature" describes the use of a template in cutting eye sockets.
1889 Jumeau claims to employ 1,000 people and to produce 300,000 dolls a year.
1889 Eiffel Tower doll game published for the American market.
1891 The Bee Mark (No.36987) registered.
1892 Coloured dolls introduced.
1893 Bébé Phonographe offered for Christmas.
1894 Leo Claretie. Publication of "Jouets: Histoire et Fabrication". With a long account
of the Jumeau factory at work.
1895 Bébé Marcheur registered.
1896 Bébé Française registered. Box lid mentions that natural hair is used in addition to mohair and sheepskin.
1897 Carrier-Belleuse mentioned as designer of head.
1899 Societe Française de Fabrication de Bébés et Jouets formed. The registered office given as the Jumeau address at 8, Rue Pastourelle.
1900 Jumeau dolls displayed as part of the S.F.B.J. stand at the Paris Exposition.

BIBLIOGRAPHY

D'Allemagne. Henry René. Histoire des Jouets. Hachette & Cie. 1902
Antwerp. Universal Exhibition 1887. Official Report.
Bartley, G.C.T. Illustrated London News. November 1867.
Bicknell, Anna L. Life in the Tuileries under the Second Empire. Fisher & Unwin 1895.
Calmettes, Pierre. Les Joujoux. Gaston Doin. 1924.
Capia, Robert. Les Poupées Françaises. Hachette. 1979.
Claretie, Leo. Les Jouets. Ançienne Maison Quantin. Librairies-imprimeries Reunis. 1894.
Cuzacq, P. La Naissance Le Marriage et le Décès. Honore Champion. 1904.
Coleman. D, E & E. The Collectors Book of Dolls Clothes. Robert Hale 1976.
Dupeux, Georges. French Society 1789-1950. Methuen & Co. 1976.
Edwards, Bentham M. Twentieth Century France. Chapman & Hall. 1917.
Fournier, Edouard. Histoire des Jouets et des jeux d'enfants. Dentu ed. 1889.
Franklin, Alfred. La Vie Privée d'Autrefois. L'Enfant. E. Plon. Nourrit et Cie. Paris 1896.
Guerard, Albert Leon. French Civilization in the Nineteenth Century. Fisher & Unwin. 1914.
Hanotaux, Gabriel. Contemporary France. 3 vols. Constable & Co. 1903.
Hillebrand, Karl. France and the French in the Second Half of the Nineteenth century. Trubner & Co. 1881.
Hueffer, Oliver Maddox. French France. Ernest Benn 1929.
London. Great Exhibition 1851. Jury reports and list of awards. 1852.
Londres. Exposition Universelle 1862. Section Française. Catalogue Officiel. Paris. Imprimerie Imperiale. 1862.
London. International Exhibition 1862. Reports by Juries. William Clowes & Sons. 1863.
Lynch, Hannah. French Life in Town and Country. George Newnes. 1901.
Menpes, Mortimer and Dorothy. Worlds Children. Adam and Charles Black. 1903.
Paris Exposition 1855. Official English Guide. Gervais & Co. Paris.
Paris Exposition Universal 1855. Official Report. 1855.
Paris. Travaux de la Commisssion Française. Imprimerie Imperiale. 1855.
Paris Exposition 1867. Report of Artisans. 1867.
Paris Universal Exhibition 1867. Complete Official Catalogue. English Edition. J.M. Johnson & Sons. 1867.
Paris. Exposition Universelle Internationale. Dec. 1878.
Paris. Official Catalogue. Liste des Récompenses. Class 42. 1878.
Paris Exposition 1889. Report Class 40.
Paris. Rapports du Jury International Paris Exposition. 1889. Ministère du Commerce 1891.
Paris. Rapport du Jury International. Jeux et Jouets. Class 100. Imprimerie Nationale 1902.
Paris. Musée de Costume. Au Temps des Petites filles modèles. Musée de Costume. Paris 1958-9.
Porot, Dr. Jacques. Article. Jumeau bodies. International Toy and Doll Collector. No.3.
Pratz, Claire de. France from within. Hodder & Stoughton. 1912.
Rhodes, Albert. Monsieur at Home. Field & Tuer. Undated, circa 1870.
Stoeckl, Baroness de & Edwards, Wilfred S. When men had time to love. John Murray 1953.
Tissandier, Gaston. Article in La Nature 1888.
Uzanne. Les Modes de Paris. Societe Française d'edition d'Art. 1898.
Vandam, Albert D. French Men and French manners. Chapman & Hall. 1895.
Vasili, Count Paul. Society in Paris. Chatto & Windus. 1890.
Vizetelly, E.A. The Court of the Tuilleries 1852-1870. Chatto & Windus. 1907.
Waddington, Mary King. Chateau and Country Life in France. Smith Elder & Co. 1908.
Whitton, Margaret. The Jumeau Doll. Dover Publications 1980.
Whitehurst, Felix M. Court and Social life in France under Napoleon 111. Tinsley Bros. 1873.

ndex

igures in italics refer to illustrations

American Price Guide

Dollar prices are kindly supplied by Countess Maria Tarnowska, international dealer in antique dolls who exhibits at leading shows in America. The valuations are approximate and assume complete originality and near perfection.

Page	Price	Page	Price	Page	Price	Page	Price
13	2,800	41	5,200	65	4,200	87	8,000
15	15,000	43	4,000	67	2,250	89	2,000
17	15,000	45	3,000	69	2,900	91 left	1,600
19	4,800	47	4,300	71	3,300	right	2,800
21	2,700	49 left	2,000	73 left	3,500	93	3,500
23	5,200	right	3,200	right	3,500	95 standing	3,800
25	5,500	51	7,500	75	3,200	seated	4,000
27	8,000	53	12,000	77 standing	4,600	97	1,600
29	4,600	55	2,000	seated	4,400	99	3,500
31	4,800	57	4,500	79	3,400	101	2,900
33	6,000	59	3,400	81 larger	3,000	103 Very rare. Not enough	
35	4,000	61	5,600	smaller	1,500	samples to estimate.	
37	6,500	62 left	3,300	83	4,400	105	5,200
39	15,000	right	3,400	85	4,500	107	3,700

It should be noted that the doll market is highly volatile and there are sometimes changes in fashion that affect prices.

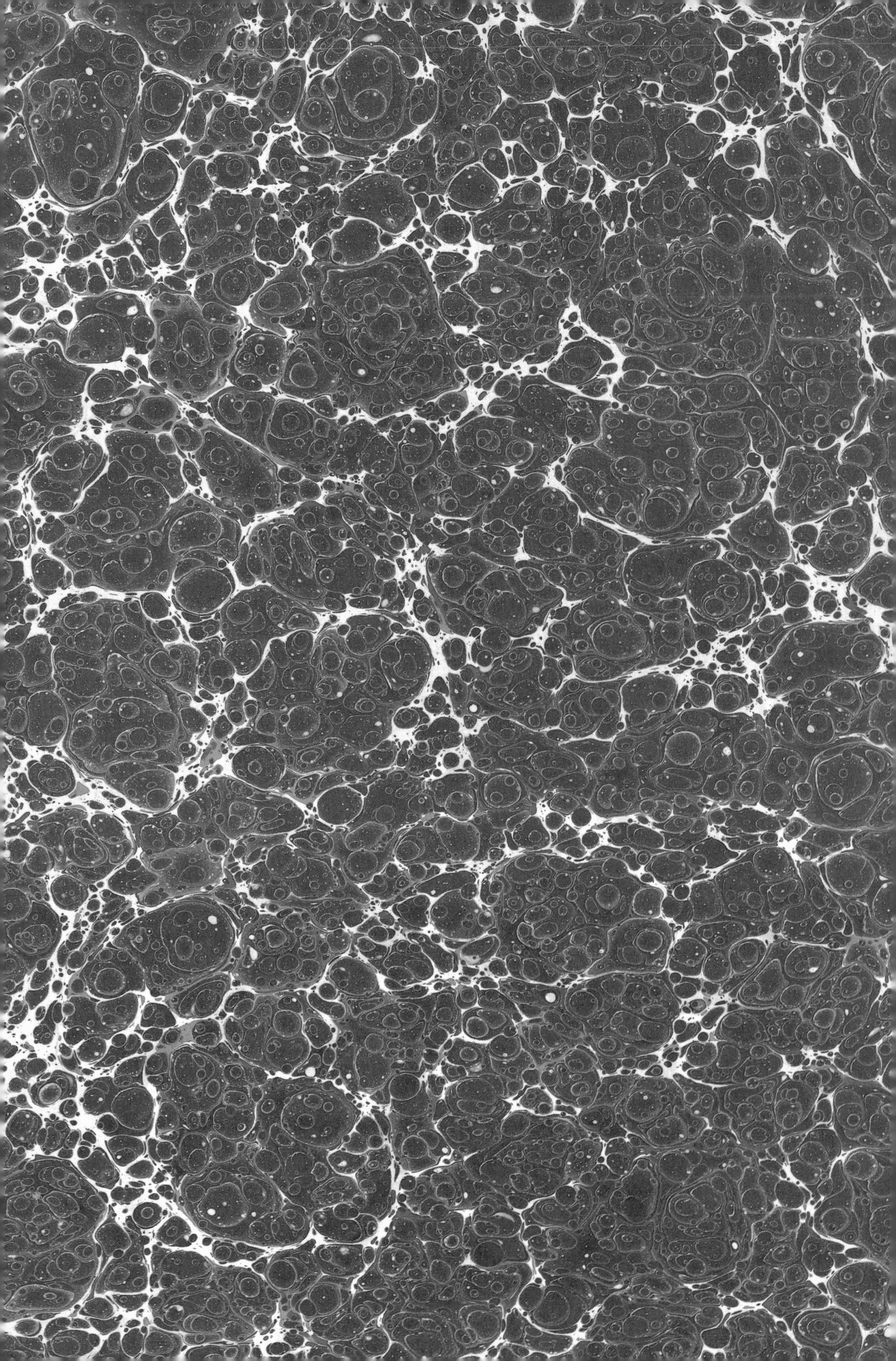